Fighting for the River

Fighting for the River

GENDER, BODY, AND AGENCY IN
ENVIRONMENTAL STRUGGLES

Özge Yaka

UNIVERSITY OF CALIFORNIA PRESS

University of California Press
Oakland, California

© 2023 by Özge Yaka

Cataloging-in-Publication Data is on file at the Library of Congress.

ISBN 978-0-520-39360-8 (cloth : alk. paper)
ISBN 978-0-520-39361-5 (pbk. : alk. paper)
ISBN 978-0-520-39362-2 (ebook)

Manufactured in the United States of America

32 31 30 29 28 27 26 25 24 23
10 9 8 7 6 5 4 3 2 1

To Rosa Mavi

Contents

Illustrations

Abbreviations

AKP	Justice and Development Party (*Adalet ve Kalkınma Partisi*)
CBR	Central Black Sea Region
EAR	East Anatolia Region
EBR	East Black Sea Region
HEPP	hydroelectric power plant
MR	Mediterranean Region
WBR	West Black Sea Region

Acknowledgments

This book has been a long time in the making. It is the culmination of the research I started in 2013 as a postdoctoral fellow at the Graduate School of North American Studies (GSNAS), Freie Universität Berlin. I was a part of the *Rethinking Crisis* project, which was led by Nancy Fraser and supported by the Einstein Foundation Berlin. Thus, the first person I need to acknowledge here is Nancy for choosing my proposal out of hundreds of others—a proposal written by a young, unknown woman working at a provincial university in Turkey. "I am more interested in good ideas than impressive CVs," she told me later, and she supported me and my work fully from beginning to end.

I spent two and a half years as a member of the project team, working alongside fantastic team members (Blair Taylor, Ethan Zane Miller, and Larry Reynolds), and then working as a guest professor at the GSNAS, completely free to engage with my work and with all the interesting things happening in Berlin at the time. It was our own Weimar years: we were young(er) and so much was happening in Berlin. We discussed interesting things with interesting people and lived the best version of the academic life. In addition to Ethan, Blair, and Larry, I want to thank Margit Mayer for being who she is, David Bosold for making GSNAS a welcoming and supportive place, and the Einstein Foundation for financing our project(s).

It was during that time that I conducted most of the field research that is at the heart of this book. I would like to thank Erdem Evren, Cemil Aksu, Ercan Ayboğa, Bora Sarıca, Kamil Ustabaş, Kamile Kaya, Mehmet Gürkan, Mustafa Koç, and Ahmet Öztürk for sharing their invaluable knowledge and experience with me and helping me establish connections in the field. I am grateful to Onur, Medet, Şenol, Mehmet Başar, Abdullah Amca, Seniye, Esma, Sultan Teyze, and many others who guided and hosted me and traveled with me during my journey through the villages and valleys of the country. I am indebted to more than a hundred people who trusted me to share their stories and their struggles. I only wish I could have done more to repay their trust.

After my Berlin years, I spent a year in Paris at the Collège d'études mondiales, FMSH-Paris, again with Nancy Fraser, as a Gerda Henkel Postdoctoral Fellow under the auspices of Nancy's *Rethinking Global Justice* chair. That year was crucial for my intellectual development, as it was during that time that I engaged with feminist theory, environmental justice literature, and phenomenology. The fellowship (thanks to Nancy, the Gerda Henkel Foundation, and the Collège d'études mondiales) gave me the time and space to develop my body-centered, feminist phenomenological framework. Paris, of course, was beautiful, and the friendship of Eda Çelik, Serdar Çelik, and Stephanie Alexander made it a memorable year. Thanks also to Sara Guindani for her warm, friendly, and supportive attitude during my time at the Collège.

After my return to Berlin, I gave birth to my daughter, Rosa Mavi, and decided to devote my time to her until she started going to day care. After one and a half years of intensive motherhood, I was awarded an Alexander von Humboldt fellowship, which helped me reenter academic life. Sincere thanks go to my generous host, Jürgen Mackert, and the Center for Citizenship, Social Pluralism, and Religious Diversity at the University of Potsdam. The fellowship gave me the opportunity to be a part of academic life (teaching, attending and organizing meetings, lecture series, etc.), but also allowed me to focus on my own work, especially on writing this book. I am grateful to Jürgen, as well as to my colleagues at the Center: Zafer

Yılmaz, Muzaffer Kaya, Hannah Wolf, Felix Lang, Christian Schmidt-Wellenburg, and Max Oliver Schmidt.

I enjoyed a fair amount of institutional and collegial support during the years in which I developed my framework (a rather unusual one for the study of environmental movements) and worked on my manuscript, but for the most part I was intellectually on my own, not being part of an established network, school of thought, or even a well-defined discipline! Hence, it was very important for me that the women I look up to—scholars of feminist, phenomenological, and environmental theory—encouraged me when I shared my work with them. For this I am particularly grateful to Diana Coole, Ingrid Leman Stefanovic, and Stacy Alaimo. Their encouragement was crucial for helping me to develop the courage, self-determination, and confidence to pursue my own intellectual path.

Serhat Karakayali, Erdem Evren, Nur Yasemin Ural, Firoozeh Farvardin, Aseela Haque, and Laura Matt read and commented on various chapters of this book. Erdem, Yasemin, and Firoozeh (and Nader) are also my closest friends in Berlin. Berlin became my home, thanks to them. Thanks also to my friends in Turkey, especially to Çağdaş, Burcu, and Duygu for always being there for me, even from a distance. My parents have always been my safety net, and they still are. They make me feel that I will always have a home—or another home—in one of the most beautiful parts of the world, amid orchards of orange, olive, and almond trees. I feel that I have the most tempting plan B in life.

Coming back to the book, I am grateful to the anonymous reviewers for their deep engagement with the manuscript and constructive comments. My editors at the UC Press—Stacy Eisenstark, Chloe Layman, and Chad Attenborough—have been wonderfully supportive. Brian Donahoe contributed to the book substantially with his elaborate language editing, which I desperately needed.

I have written this book under (relatively) difficult circumstances. In the last six years, I became a mother and survived a law case (the Academics for Peace case in Turkey) and a pandemic. My partner, Serhat, supported me unfailingly throughout this entire time. We have built a life together and raised our daughter. He took over most of the household and childcare responsibilities during the pandemic (how wonderful it is that this time a woman can write that about a man), giving me the space and time I needed to focus on my book. He even learned a cartography program to draw the maps in the book. I feel incredibly lucky to be able to share my life with him. Our wonderful babysitters at the time, Duygu and Özgür, a psychologist and artist respectively, provided indispensable help with Rosa Mavi during the pandemic when our day care center was closed. A heartfelt thanks to both of them.

And finally, to my daughter Rosa Mavi, who gave me a hard time with her cute sexism ("Working is not good for mothers!") and strict supervision ("Haven't you finished writing this book of yours?!"). My life might be challenging, raising her with my partner without any local family support and trying to build a career under far-from-ideal circumstances, but it has all become so much more precious since I gave birth to Rosa. Her sheer existence makes me crazy happy. This book is dedicated to her.

Introduction

GENDER, BODY, AND RELATIONALITY IN
THE STRUGGLE FOR THE ENVIRONMENTAL COMMONS

We, as women, won't allow them to construct a hydropower
plant here. We don't even count on men. . . . Bring them on,
if they dare, if any brave fellows think they can come here
. . . let them try. We will cut them to pieces. We know how
to use guns as well. We take the risk. They really shouldn't
force us. Don't make people go mad.

Selime, a middle-aged woman from the village of Arılı (East
Black Sea Region—EBR)

Selime is one of the countless women who fight against private, small-scale
hydroelectric power plants (HEPPs) all over Turkey. The struggles for and
around environmental entities, rivers in this case, are defined and discussed
in the fields of political ecology, social movement studies, environmental
justice, and environmental anthropology. In close dialogue with these bodies of literature, this book seeks to develop a novel, body-centered perspective on grassroots environmental movements in local communities, paying
particular attention to gender as an integral aspect of our constitutive relation to nonhuman entities and environments. The body-centered perspective of this book is based primarily on an empirically grounded reading of
Maurice Merleau-Ponty and his followers. But I put Merleau-Ponty's work
and contemporary (critical, feminist, post, and eco-) phenomenology in
dialogue with broader critical and feminist theories, environmental humanities, human geography, posthumanist and new materialist perspectives,
and Indigenous studies. The novel framework of the book emerged, slowly
but surely, through these multi- and transdisciplinary engagements and

1

dialogues, anchored in profoundly empirical research, which enabled me to develop my body-centered approach into a comprehensive framework. Operationalizing this framework, I not only analyze the anti-HEPP activism in Turkey with a focus on the corporeal—that is, sensory and affective—experience of and interaction with river waters; I also develop a conceptual apparatus, encompassing discussions of gender, place, memory, identity, ontology, cosmology, ethics, and justice, that brings together the multiple dimensions of our relationship with and relationality to nonhuman entities and environments.

Hence, throughout this book I employ the body as a key methodological anchor not just to study lived experience, but to reveal how lived experience connects us to nonhuman entities and environments and to conceptualize the implications of this corporeal connectivity. *Fighting for the River* thus makes a double intervention, introducing and operationalizing a body-centered framework to study local environmental struggles, and framing them as *struggles for coexistence* (Escobar 2011; Larsen and Johnson 2017) to claim and protect the place-based, socio-ecological relations that involve nonhuman/environmental entities. Toward the end of the book, I develop the notion of *socio-ecological justice,* which frames human-nonhuman relationality as a matter of justice. My discussion of relational ontologies, ethics, and socio-ecological justice demonstrates the potential of a body-centered, phenomenological, and relational approach for rethinking our fundamental notions of living together with human and nonhuman others on Earth. *Fighting for the River* is, more than anything, an empirically grounded conceptual attempt to rethink our intimate (and gendered), social, corporeal, and affective relations with nonhuman entities and environments, and their implications for our socio-ecological existence, our identities, agencies, notions of justice, struggles, everyday practices, and ethical conduct—in short, for the future of our common life on Earth.

THE EMPIRICAL CASE AND THE EMERGENT ARGUMENT: FROM METHODOLOGY TO CONCEPTUAL FRAMEWORK

Turkey already has a relatively long history of constructing dams, especially in the context of the Southeast Anatolia Project (*Güneydoğu*

Anadolu Projesi, or GAP), which was initiated as a regional development project involving multiple dams in the Tigris and Euphrates river basins.[1] However, with the exception of a few highly controversial mega dams such as Ilısu (Eberlein et al. 2010; Hommes, Boelens, and Maat 2016) and Yusufeli (Evren 2014), the new wave of hydropower development is dominated by private, small-scale, run-of-the-river HEPPs, which require little or no water storage.[2] They have emerged as a central element of the construction and energy-based growth strategy of the ruling Justice and Development Party (*Adalet ve Kalkınma Partisi,* AKP), the so-called bulldozer neoliberalism (Lovering and Türkmen 2011; see also Erensü 2018; Evren 2022) that has held sway over the last two decades.[3] HEPP licenses and construction boomed in the 2000s after the establishment of an open energy market,[4] as private construction and energy companies were given "extraordinary latitude to evict villagers, expropriate private land, clear state forests and steamroller normal planning restrictions to meet the target of four thousand hydroelectric schemes by 2023" (Gibbons and Moore 2011).[5]

Besides destroying the country's river ecosystems and natural habitats,[6] private HEPPs also dispossess countless riverside communities that, for all intents and purposes, lose access to their rivers when they are diverted from their streambeds for a number of kilometers.[7] These communities began to organize themselves, mostly in the rural parts of Anatolia, as the effects of the first wave of plants became clear, especially around 2008 and 2009. A large and heterogeneous movement appeared in the villages and valleys of the country, which slowly led to regional and national networks (Hamsici 2010; Aksu, Erensü, and Evren 2016). The anti-HEPP movement in Turkey is a facet of a broader resistance to extractivism, the private enclosure of environmental commons, and the increasing exposure of natural entities and environments to the growth-and/or profit-oriented instrumental rationalities of states and markets.[8]

The transgressive, intersectional character of contemporary grassroots environmentalisms—as they cut across issues of ecology, economy, culture, and politics—was my starting point in researching local community movements against HEPPs in Turkey. The project's focus, however, changed drastically during the research process. The empirical focus and the conceptual framework evolved in a recursive relationship to each

Map 1. The provinces of Turkey that are mentioned in the book: Ordu, Giresun, Trabzon, Rize, and Artvin in the East Black Sea Region (EBR); Kastamonu in the West Black Sea Region (WBR); Amasya in the Central Black Sea Region (CBR); Muğla, Antalya, and Mersin in the Mediterranean Region (MR); Erzurum and Dersim in the East Anatolia Region (EAR); and Diyarbakır in the Southeast Anatolia Region. See the appendix for detailed geographical information. Map by Serhat Karakayali.

other, in close dialogue, with the empirical taking more of the lead in shaping the conceptual. Hence this book is the result of a methodological and intellectual journey in which the empirical (data) was given the space to "speak" and, ultimately, to lead the research process toward novel, and often unexpected, conceptual engagements.

It all started with extensive ethnographic fieldwork conducted from mid-2013 to late 2015 in three regions of Turkey where hydropower development is concentrated—the East Black Sea, the Mediterranean, and East and Southeast Anatolia (the latter being Turkey's predominantly Kurdish region). To develop a comparative perspective, I visited and stayed in several provinces in each region, and multiple districts, villages, and valleys within each province. In designing my fieldwork, I used criteria such as media coverage, public visibility, and availability of local contacts to select places where there was or had been strong community resistance to HEPPs. What shifted both the focus of the study and my conceptual framework entirely, however, was the voices of the villagers themselves. I did indeed interview local, regional, and national activists,[9] as well as lawyers and academics who work on the HEPP issue and/or are engaged in the struggle, in several locations: Ankara and Istanbul; Trabzon, Antalya, and Diyarbakır (Kurdish: Amed) (the main cities of the three regions I studied); and in various district centers. I did not, however, limit my interviews to activists and intellectuals, which is the underlying reason for the "elite bias" (Benford 1997) in social movement studies. The data set I relied on most heavily is composed of ethnographic interviews (Skinner 2012) and recorded conversations with more than one hundred villagers, slightly more women than men, as well as my own observations gathered through immersive fieldwork in the various locations.

I have thus combined the qualitative methods of social movement research (Klandermans and Staggenborg 2002; della Porta 2014)—in-depth interviews with activists and participants, documentary analysis of the visual and written texts produced by the movement and of media accounts of the movement[10]—with ethnographic research methods such as fieldwork, participant observation, recorded field notes, and ethnographic interviews. I prioritized talking to villagers who live under the immediate threat of HEPP projects in order to differentiate their discourses, narratives, and political agencies, along with the processes, experiences, and relations that shape them, from the motivations of the movement activists.[11] Still, I cannot claim that the empirical data I have collected are fully void of "elite bias," as my entrée to many villages was through local activists. To balance the (in my case inescapable) bias that comes with being introduced in the field by well-known figures, in addition to triangulating multiple methods (Snow and Trom 2002) as detailed above, I talked to as many villagers as I could, when and wherever it was possible, in more formal interview settings as well as in less formal contexts (coffee houses, terraces, doorways, etc.).

This was also a strategy that allowed women's voices to be audible and to come to the fore. I knew that women were very present and active on the ground, publicly visible at demonstrations and protests in their traditional clothes, giving the whole movement a face and a voice. I also knew that women were behind the radicalism of the movement. As manifested in the words of Selime above, they were the ones who framed the struggle as an issue of life and death; they were the ones, especially if they were middle-aged or older, who talked openly about beating, killing, and being killed for the cause, whereas men used a more cautious language. Despite all that, women were severely underrepresented in the movement organizations, including local ones, as is the case in many other grassroots environmental and/or justice movements (see, e.g., Di Chiro 1992; Brown and Ferguson 1995; Kurtz 2007; Buckingham and Kulcur 2009).[12] Hence, the need to talk to at least as many women as men was clear to me, not because I went into the research with a consistent feminist methodology, but because I sensed that my research would otherwise misrepresent the movement.

Including women's voices proved to be a challenge, though, as men were more willing to talk and accustomed to taking charge,[13] while many women, despite being very confident, bold, and committed to action, tended to look to men when it came to responding to questions and giving interviews. Luckily, my insistence encouraged them, and being a woman made it possible for me to talk to them in their homes, on their doorsteps, and in the fields, during the routine flow of their daily lives, either alone or among their female peers. As a young (-looking) woman who did not fit their image of an academic (*hoca*, as we say in Turkish, a gender-neutral term that literally means "teacher/preacher" and is used for imams, school teachers, and university staff), I had the opportunity to be perceived as less "official."[14] People often asked me in the villages, "What kind of a *hoca* are you? You look like a student. You should be wearing something proper, like a *döpiyes*," a word that comes from the French *deux-pièces* ("two-piece suit") and refers to a famous style concept strictly associated with female civil servants. My less than official appearance, I believe, helped me to keep interviews informal and conversational. I also kept the conversation two-sided, which meant that I shared my own motives for researching their anti-HEPP activism if I was asked to. I tried to talk about myself as openly as I could, especially as I was asking them to share their experiences, motivations, and stories with me. This informal and conversational mode relieved the pressure of being interviewed and made it easier for women to talk openly.

As I talked to women, it became immediately clear that the discourses, narratives, and forms of action they employed to communicate their resistance against HEPPs differed substantially from those of men (Yaka 2019a). For example, during my very first field trip in the summer of 2013, after spending two and a half hours in a village coffee house in Arılı (EBR) listening to men's theories on global warming, global struggles over fresh water, imperialist plans of the United States, Israel's efforts to grab "our" waters, and the close affinity between the War of Independence and the anti-HEPP movement (protecting the country—protecting the water),[15] I talked to women in the same village and heard a completely different story. Women told me about their childhood memories of rivers,[16] their identification with the places that they feel are defined by the river's course, and more often than not, their bodily (sensory and affective) connections with the rivers' waters. They talked about growing up by the river, waking up to the sight of the river every day, and falling asleep to the sound of it every night. They talked about the sensations and bodily affects, about the joy, rejuvenation, and relaxation they felt when they were immersed in river waters after working in the fields. They talked about the memories of their parents by the river and the sight of their children and grandchildren playing in the same waters they once played in.

The centrality of memories, past and present sensations, affective responses, and the emotions generated through the corporeal connection between bodies of women and bodies of water infiltrated their narratives of the anti-HEPP movement. Those narratives reveal the interconnectedness of female bodies with river waters as the main source of women's radical opposition to hydropower. It was those narratives of women and their everyday, corporeal experiences of river waters that led me to investigate the feminist literature on body, experience, and agency. As a result, corporeal feminism (see Grosz 1994) became my entry point to the long journey of establishing my own framework. In the coming sections I will unfold, step by step, the development of this conceptual framework and the arguments that have emerged through my rigorous empirical engagement and the multidisciplinary dialogues this engagement led me into.

THE BODY-SUBJECT: CRITICAL AND FEMINIST PHENOMENOLOGY IN THE ANALYSIS OF THE GENDER-BODY-AGENCY RELATIONSHIP

Diana Coole (2005, 131), a leading feminist scholar of phenomenology and new materialism, states that phenomenologists "might begin by explaining how problems that motivate political agents originate in somatic experience." At the beginning of my research, I was not a phenomenologist, nor did I follow Coole's suggestion. It was the empirical, ethnographic data I have collected, especially my conversations with women, that forced me to search for a conceptual language that could connect their somatic experiences with their political agency. It became immediately clear to me that the Foucauldian post-structuralist literature, which treats the body as a surface on which power and discourse act and are inscribed, as a sort of object that can be observed, shaped, monitored, disciplined, utilized and deployed, is of little use to me. My work rather called for an understanding of the body not only as formed, but also as simultaneously formative (see Coole and Frost 2010). I became interested in the body not only as an object and tool of subjection, but as a vehicle of subjectivization. In the case of the anti-HEPP struggles, my interest lies in how the corporeal connection, established with river waters through routine, habitual interactions within a more-than-human lifeworld, conditions women's resistance to HEPPs. This requires attending to the body's perceptive and affective capacities through which we inhabit, experience, and act upon the world we live in.

In his recent book *Resonance: A Sociology of Our Relationship to the World*, Hartmut Rosa states that the body not only oscillates between two poles—"the body as self" and "the body as world"—it is "the constitutive basis of both" (Rosa 2019, 84–85).[17] It is this constitutive character—of perception, experience, knowledge, consciousness, subjectivity, and political agency, as well as of our connectivity with others and with the more-than-human world we inhabit—that makes the body central for the purposes of this book. Attending to the body both as self and as world, as "the constitutive basis of both," requires understanding its perceptive and agential capacities as immersed in the power-laden materiality of the world. Developing this particular understanding of the body, I followed in the footsteps of feminist scholars—corporeal feminists, as Grosz (1994) would call them—who were inspired by or engaged in either Spinozan/ Deleuzian or phenomenological traditions and presaged contemporary new materialist and posthumanist feminisms in many different ways.[18] My sympathies for the Spinozan/Deleuzian tradition aside, the particular advantage of phenomenology, especially the work of Merleau-Ponty, is that it provides conceptual tools for exploring women's concrete, lived experiences and embodied subjectivities as shaped within a more-than-human world, in the context of their environmental activism. As an empirical parenthesis here I would like to point out that while *Fighting for the River* draws mostly on women's experiences and narratives in building its main arguments, it does not exclude men's.[19] The need to include both women's and men's experiences and perspectives is based on an understanding of gender as inherently relational, and this of necessity includes the relation between sexes (Scott 1986, 2010). Men's experiences and narratives are not just used to reveal gender differences; they are also used to complement women's, as those differences are not essential but shaped by "the social positionings of lived bodies" (Young 2002, 422) along the lines of power hierarchies, cultural norms, and spatial structures.

What makes Merleau-Ponty's work particularly useful for the purposes of this study is his differentiation between the lived/phenomenal body and the physical/biological body. The body-subject is firmly based on the former, as "historical, social, cultural weavings" of materiality (Grosz 1994, 12). The concept of the body-subject analytically helps to maintain the subject as embodied and situated, while underpinning the analysis of experience and agency as emergent within a world of bodily encounters. Body-subject is situated, embodied, and enmeshed in a world of power-laden materialities, but still experiences the world from a particular location (of a particular body; see Rich 1984), and acts as a part of it. Merleau-Ponty located the body-subject within a relational ontological framework, especially in his later works, but he never dissolved it altogether. This makes his phenomenological approach particularly useful for a critical feminist analysis that recognizes that not only our physical bodies, but also our social beings and agential potentialities emerge within a more-than-human world of encounters, but that could not afford to give up on the notions of political agency and subjectivity altogether.

It is the task of critical feminist studies to flesh out the body-subject as differently sexed, raced, aged, as spatially and historically located, and as embedded within specific social, cultural, and sexual relations of power. The task is to attend to "embodied, situated and often more affective forms" (Simonsen and Koefoed 2020, 9) of first-person experience, as situated in an intersubjective, intercorporeal, and more-than-human interworld.[20] This book demonstrates a certain way of fulfilling this task by diving deep into empirical work and coming to the surface with a conceptual framework. It shows, on the empirical level, how to operationalize the body-subject, how our lived bodies connect us to our human and non-human environments, how we weave a bodily web of human and nonhuman beings through our routine, everyday interactions with them, and how this human-nonhuman web of relations defines our lifeworld, sociality, and agency. It illustrates how the physical, perceptive, and affective capacities of the lived body, as shaped by the socio-spatial organization of everyday practices, and historically specific constellations of social, cultural, and sexual relations, function as vehicles of agency, subjectivity, and relationality.

Framing bodily senses and affects as media of embodied subjectivity and human–nonhuman relationality (see below), in this book I transmit the lived (phenomenal) body from its familiar terrain of practice, experience, and identity to unexpected territories of political agency, memory, place, relational ontology, and more-than-human justice. I not only develop a novel, body-centered approach to studying grassroots environmentalism and the relationship among gender, environment, and activism, advancing the existing frameworks in related fields such as political ecology, human geography, social movement studies, and environmental justice; my arguments also contribute to phenomenological theory, especially to its feminist, critical, and post- currents, by discussing, modifying, and redeveloping concepts such as phenomenological reduction (bracketing), the anonymous body, the phantom limb, the *écart*, *Einfühlung*, *Ineinander*, and, of course, the lived body and the flesh, using the formative potentials of empirical work. In this book I go beyond merely "employing" the concepts to analyze the empirical; I put them in a close, mutually constructive, and transformative dialogue with one another.

RESOURCE VS. LIFEWORLD: BODIES, NATURES, ENVIRONMENTS

> Any elementary analysis of the ways in which human beings
> come to relate to the world, experience and perceive it, act and
> orient themselves within it, cannot but begin with the body.
> Hartmut Rosa, *Resonance: A Sociology of Our*
> *Relationship to the World*

It was during my first field trip to the East Black Sea Region that I clearly identified gendered differences (e.g., in the ways men and women frame their grievances, express their motivations, and enact their political agencies) and the centrality of women's everyday corporeal experiences of river waters in the formation of their political agency.[21] The same field trip also falsified a key assumption in the literature, that is, that rural communities' struggles for the environmental commons—which are mostly discussed under the category of "natural resources" in the literature—are driven by their immediate economic dependence on the commons they fight for. Hence the articulation of these struggles as "environmentalism of the poor" (Martinez-Alier 2002), "resource conflicts," and "resource rebellions" of the "ecosystem people" (Nixon 2011). In the northern parts of the East Black Sea Region, however, where HEPP projects are concentrated[22] and the resistance against them is at its strongest (the region is known as the stronghold of anti-HEPP resistance),[23] river waters are used neither for agriculture nor in the household, as the rainfall alone sustains the monocultural tea cultivation (and hazelnut in the western parts of the region).

In fact, the anti-HEPP struggle in the East Black Sea Region was not a typical "resource rebellion" in the sense of rural communities struggling to protect immediate livelihoods, as in the Mediterranean Region, nor was it motivated by claims of recognition, as in the Kurdish parts of the country (see chapter 1). Standing out as the atypical case among the three regions where I have conducted research, the East Black Sea Region emerged as the empirical focus of my study. Another empirical parenthesis is in order here: despite focusing on the East Black Sea case, I have continued to work on the cases of the Mediterranean and East and Southeast Anatolian Regions (the main Kurdish regions). These two other cases have proven to be important for the purposes of the study, as the comparative dimension has helped me, on the one hand, to understand and interpret the dynamics of the anti-HEPP struggle from a broader perspective, as is clearly visible in chapters 1 and 4, and, on the other hand, to clarify my argument, which draws most heavily on the East Black Sea case.

The case of the East Black Sea Region demonstrates, empirically, that the connection between resisting communities and river waters cannot be reduced to an instrumental relationship, as the language of "resources" implies. This connection is established, instead, through bodily senses and affects, through memories, histories, and emotions, through the joint place-making agencies of human communities and river waters. Hence my empirical focus on the case of the East Black Sea Region forced me to go beyond conventional political ecology, which is marked by the language of "resources" and livelihoods. It led me toward another vocabulary, not to ignore the relevance of livelihoods in driving many local environmental struggles, but to reveal its limits in translating many other struggles that are motivated by everyday experiences, human-nonhuman interactions, and corporeal and affective connections. It is peculiar that the bodily nature of our relationship with our natural environments remains largely unexplored and is seen as somehow irrelevant to our struggles to protect those environments. It is in fact this bodily relationship, established mainly through senses and affects, that conditions our behavior toward our environments and shapes our political agencies, as manifested in the narratives of the East Black Sea women regarding their corporeal connection to river waters.

To explore the relevance of this bodily, experiential relationship to our behaviors, notions, claims, and struggles regarding our immediate environments, I recovered the concept of *lifeworld* from the phenomenological geography tradition (Buttimer 1976; Seamon 1979). Lifeworld, a concept that originates in Husserl's work, denotes the routine, familiar, and taken-for-granted world of everyday experience (see chapter 2). David Abram (2017, 40) describes it as "the world of the clouds overhead and the ground underfoot." Phenomenology is, first and foremost, a methodology for revealing these taken-for-granted aspects of our lived experience that are rendered invisible, a call to go *züruck zu den Sachen selbst* ("back to the things themselves"), as Husserl put it. Our everyday, routine interactions with our environments, with the nonhuman entities and things, are among these taken-for-granted aspects. In the case of the East Black Sea Region, everyday interactions with river waters—seeing, hearing, touching the river waters; working, playing, relaxing, socializing by the rivers, and sometimes in them; falling in love next to the rivers, and sometimes with them—were taken for granted until the HEPPs functioned as a violent form of phenomenological reduction that forces people to go "back to the things themselves," especially to the rivers, making these interactions and experiences visible in the face of losing them (see chapter 3).

These experiences and interactions with the river waters are not immediately relevant to people's livelihoods, but they are central to their lifeworlds. Rivers are central elements of cultural, aesthetic, and affective geographies in the region—as in many other places in the world. Life in the villages and valleys of the East Black Sea Region is a practice of living together with and alongside the rivers. Drawing on the case of anti-HEPP struggles in the region, one could say that the position of an environmental entity such as a river within the lifeworld of a person or a community is crucial to the formation of the political agency necessary to protect that entity. What drives the resistance in this particular case is the centrality of rivers to the sensory, affective, and emotional world of experience and to the making of places, histories, memories, and heritage. This understanding of environmental entities not as "resources" but as constitutive parts of the lifeworld resists the subordination of environmental entities, the commons where they are located, and the social-affective relations established with and around them, to the profit- and growth-oriented logics of states and markets. In the case of private HEPP development in Turkey, what the local communities resist is the very praxis of treating the rivers as mere commodities, the waters of which can be traded between the state and private companies. The term *God's river,* which is commonly used across different communities in their struggles against HEPPs, is a manifestation of such resistance, as I explore in chapter 1.

The concepts of the lived body and the lifeworld are central to my conceptualizations of place, memory, and identity, inspired by phenomenology and indigenous studies. Understanding the place in relation to the lived experience that involves our relationship with human and nonhuman beings (see Larsen and Johnson 2012b), and recognizing the central importance of nonhuman entities, in our case the rivers, to the *sense of place* (see Malpas 2018), I relate place-bound identities, which motivate place-based struggles, to our daily engagements and embodied encounters with human and nonhuman others in a given place. Through an extensive discussion of body and place memory (see chapter 4), I demonstrate how the corporeal reenactment of the past, through the everyday sensory interactions with river waters, is constitutive of political subjectivities. In this sense, the rivers signify the "immanence of the past in the present" (Casey 2000, 168), as the past seeps into the present and the future with the flow of the river, as it connects, in a material and experiential manner, memories of childhood and ancestors to the experiences of children and grandchildren. This phenomenological, body-centered perspective enables me to demonstrate how central building blocks of our identities, such as places, memories, histories, and heritage, are also produced and conserved through corporeal—sensory and affective—interactions with river waters within a more-than-human interworld.

RELATIONAL ONTOLOGIES, MORE-THAN-HUMAN LIFEWORLDS, AND THE NOTION OF SOCIO-ECOLOGICAL JUSTICE

> [The body's] double belongingness to the order of the object
> and the order of the subject reveals to us quite unexpected
> relations between two orders.
> Maurice Merleau-Ponty, *The Visible and the Invisible*

The lived body is a medium of connectivity. It is the material basis that connects the self and the world. When the lived body is placed as an epistemic and methodological anchor within the nondualistic, relational ontology of the body and its environment (Merleau-Ponty 1968, 2003; see also Coole 2007; Simonsen and Koefoed 2020), it provides us with an entry point to study how the more-than-human world is encountered and experienced. Two main aspects of Merleau-Ponty's phenomenology are crucial here in operationalizing the lived body as a medium of connectivity. The first one is the embedding of the body-subject within the ontological realm of the *flesh*,[24] that is, within a generative, more-than-human interworld. It is through the flesh of the body that we became part of the flesh of the world. The body is thus embedded in a relational network of materiality, that is, the flesh of the world it shares with other bodies, organisms, and things, while simultaneously constituting singularity within this network. Understanding the body-subject as embedded in this more-than-human network of being, which lives *in* and *as* flesh, makes it possible to attend to first-person accounts and lived experiences without falling into the trap of subject-centered ontology. Thus, Merleau-Ponty preserves the body-subject while going beyond the subjectivism of early phenomenology and overcoming the ontological and foundational distinction between the subject and the object (see Merleau-Ponty 1968; Coole 2007; Butler 2015).

The flesh, as the ontological realm of being where the body-subject unfolds, precedes the subject and object as inherently reversible positionalities. But at the methodological-analytical level, the divergence between them, between the self and the world, remains. This methodological difference "between the identicals" (Merleau-Ponty 1968, 263) opens an invaluable conceptual space for the embodied subject, which can only be articulated as contingent, emergent, and sentient, but remains to exist as a unit of experience and action. It acts as an antidote to the tendency of new materialist and posthumanist approaches "to give unlocated accounts of materiality" (Clare 2019, 45) by providing a solid ground for exploring the ways in which we experience and relate to the material world in place.

The second important aspect of Merleau-Ponty's work, which is crucial for establishing the lived body as a medium of connectivity, is more related to epistemology. It is this framing of sensory perception not as a subjective experience but as "a communication or a communion," "a coupling of our body with the things" (Merleau-Ponty 2012, 334). It is through "this reciprocity, the ongoing interchange between my body and the entities that surround it" (Abram 2017, 52), that we live in constant contact and communion with the more-than-human world. I employ the lived body, in relation to the onto-epistemological conceptions of the flesh and to sensory perception as a more-than-human communion, as an analytical angle to study the relational ontologies of human-nonhuman existence and the "already existing" relational ethics (Thomas 2015) of Indigenous and place-based communities beyond culture and belief (see chapter 5).

Through this conception of the formative, agential body located within a relational ontology of human and nonhuman life, agency, and materiality, I focus on the sensory and affective capacities of the lived body in connecting us to the nonhuman world, in this case, to the river waters. I also uncover the centrality of more-than-human relations and bodily interactions to the enactment of the cosmological or religious value of environmental entities such as land and water. In this sense, I believe, in line with such scholars as Soren C. Larsen, Jay T. Johnson (Larsen and Johnson 2017), and David Abram (2017), phenomenology, especially the work of Merleau-Ponty and his followers, can serve to advance the dialogue with Indigenous relational ontologies and ethics.

If we return to the concept of the lifeworld, this relational ontological perspective requires framing it as an intercorporeal and more-than-human interworld. With reference to the German origins of the concept, this might mean relating Husserl's *lebenswelt* (lifeworld) to Uexküll's *umwelt* (environment, surroundings). Relating *lebenswelt* with *umwelt* and revealing the ways in which nonhuman entities become a part of our more-than-human lifeworld through our everyday experience of and corporeal connection to them have significant implications for environmental movements and the environmental justice literature that examines them. This requires understanding these movements, and their claims for justice, beyond the framework of "resource struggles," and rethinking our conceptions of grievances, rights, sociality, and justice in light of a relational ontological perspective. By drawing on the relationality of human and nonhuman beings not only at the physical and chemical levels, but also in social and affective spheres, such processes of rethinking aim at reconfiguring sociality and social justice in a more-than-human world. The notion of *socio-ecological justice*, which I introduce toward the end of the book, translates the ecological embeddedness of our social existence into the vocabulary of (environmental) justice. Drawing on the justice claims produced within grassroots environmental movements and highlighting the *excess* of relationality of human and nonhuman life over the established notions of justice, socio-ecological justice incorporates our transversal connection to nonhuman entities and environments into our conceptions of social and environmental justice.

THE PROMISE AND THE STRUCTURE OF THE BOOK

This book attempts to establish a novel framework for studying local environmental movements, gender and environment, and human-nonhuman relationality in the particular empirical context of anti-HEPP movements by developing a body-centered, phenomenological framework. And it is a book about local environmental movements in Turkey, which might not appear as the obvious setting for such a venture. The broader Middle East, a region discussed mostly via repressive authoritarian governments, sectarian conflicts, and war over the last few decades, is even less so.[25] I hope the conceptual ambition of the book is seen as an intervention, not an individual one, but one made in the name of a large group of young scholars from the Global South who are building innovative theories and conceptual frameworks on the foundation of rigorous empirical engagement and whose contributions go far beyond area studies.[26] As a transdisciplinary venture, *Fighting for the River* aims at contributing to, or at least being in discussion with, a variety of fields and disciplines, from feminist theory to contemporary phenomenology, from human geography to environmental humanities, from social movement studies to Indigenous studies, from affect and memory studies to water studies, from posthumanism to theories of justice, and, of course, Turkish and Middle Eastern studies.

In chapter 1, "Saving 'God's Water': Motivations and Dynamics of the Anti-HEPP Struggle," I set the stage for the main arguments of the book and provide a detailed account of the empirical context of the research: the legal and political processes surrounding private, run-of-the-river HEPP development projects and local resistance against them. Employing the literature on the commons, extraction, and expropriation, the chapter discusses the social struggles and legal battles around which the resistance against HEPPs is organized. It also identifies regional differences that shape the use of river waters and the dynamics of the struggle. Chapter 1 tells a common story of resistance against the environmental dispossession that is imposed on rural/peasant and Indigenous communities, but one that unfolds in unprecedented ways. In this sense the chapter illustrates the value of case studies for telling a common story anew, modifying and transforming it in each and every iteration by showing us the many ways in which common processes can unfold in different socio-spatial contexts.

Chapter 2, "Resources, Livelihoods, Lifeworld: Linking Gender and Environment through the Lived Body," discusses women's activism against HEPPs, especially in the empirical context of the East Black Sea Region, in dialogue with the critical feminist literature on gender and environment. Building on the strong foundations this literature has established, the chapter also responds to its limitations, especially to the analysis of embodied experience, which tends to be limited to its instrumentality in sustaining livelihoods. Returning to the founding fathers of phenomenology and the reception of their work in the field of human geography, the chapter introduces the concepts of *dwelling* and *lifeworld* in order to situate embodied experience beyond the instrumentality of livelihoods. It then goes on to lay out the main pillars of a body-centered perspective and to discuss its conceptual potential for studying relations between gender, environment, and political agency. The chapter also situates the lived body within a power-laden social world, illustrating that the body-subject is both shaped and constrained by social, cultural, and sexual relations of power while being simultaneously formed through somatic interconnections within a more-than-human world.

Chapter 3, "Sense, Affect, Emotion: Bodily Experiences of River Waters and Emergent Political Agency," deepens the book's engagement with Merleau-Ponty's work, combining it with contemporary approaches in phenomenology, affect studies, ethnographies of water, emotional geographies, and feminist theory. It draws on in-depth ethnographic data to explore the sensory and affective connections between human bodies and river waters by scrutinizing women's everyday lives and bodily experiences, which routinely involve the sight, the sound, the taste, and the touch of the river waters. The chapter contributes to phenomenological theory by operationalizing, modifying, and rethinking concepts such as phenomenological reduction, the flesh, and *Einfühlung* on the basis of extensive ethnographic data. Thus, it puts a hybrid phenomenological theory into a transformative conversation with ethnography to explore this corporeal, sensory connection between bodies of women and bodies of water, which is established through and within habitual practices, produces a series of affective and emotional attachments, and conditions political agencies to protect river waters.

Chapter 4, "Place, Body, Memory: River Waters and the Immanence of the Past in the Present," extends the scope of the book's body-centered analysis to investigate the relations between water, place, and identity, which are constitutive of embodied political subjectivities in the struggles for and around water bodies, including the anti-HEPP struggles in Turkey. In doing so, the chapter focuses on body and place memory, which connects places and environmental entities to identities, histories, and heritage. The chapter makes extensive use of empirical data, combining it with a phenomenological analysis of place and memory. It draws on Indigenous studies, critical geography, memory studies, and ethnographies of water, place, and the struggles around them to situate the lived body and the embodied nature of "being-in-place" at the center of the relationship between water, place, and identity. The chapter maintains that personal memory, family history, and cultural heritage survive through the everyday sensory relation with river waters, as sensate—haptic, aural, and visual—experiences actualize the past in the present.

Chapter 5, "Ethics, Ontology, Relationality: Grassroots Environmentalism and the Notion of Socio-Ecological Justice," explores the implications of the body-centered, phenomenological approach that has been developed throughout the book for our conceptions of social and environmental justice. Toward this aim, the chapter constructs a fruitful conversation between social theory, theories of justice, and environmental and ecological justice, on the one hand, and environmental philosophy, anthropology, Indigenous studies, critical geography, eco-phenomenology, and posthumanism on the other. Discussing empirical data through this wide range of references, the chapter frames Merleau-Ponty's relational ontology as a conceptual angle to the "already existing" (Thomas 2015) relational ontologies and ethical practices within Indigenous and rural communities beyond the frame of recognition, beyond culture and belief. It also maintains that our corporeal connectedness to nonhuman entities and environments requires rethinking our concepts of sociality and justice. By framing human-nonhuman relationality as a matter of sociality and justice and by introducing the notion of socio-ecological justice, the chapter aims to expand the borders of existing conceptions of social and environmental justice.

The concluding chapter, "Toward an Ecological Approach to Lifeworld, Sociality, and Agency," summarizes the main arguments of the book, focusing on the body-centered approach that has been developed throughout. The chapter discusses the conceptual implications of thinking politics and political agency from the perspective of the body while situating the body within a more-than-human lifeworld. Maintaining that nonhuman life and entities are integral not only to the *umwelt* (environment, surroundings), but also to the *lebenswelt* (lifeworld), the chapter underscores the constitutive role of nonhuman beings in the making of sociality, identity, and embodied subjectivity. The chapter employs concepts such as *ecosociality* (Whitmore 2018) and *ecosophy* (Guattari 2000) to bridge its body-centered, feminist, and phenomenological approach with a relational ecological conception of life, sociality, and agency.

1 Saving "God's Water"

MOTIVATIONS AND DYNAMICS OF THE ANTI-HEPP
STRUGGLE

We cannot call this a natural disaster when it is man's intervention that is the underlying cause.

Ömer Şan, spokesperson for the Brotherhood of the Rivers
Platform (DEKAP), quoted in Yackley 2020

Dereli is a small town in Turkey's Giresun Province, located in the northeastern part of the country, in the East Black Sea Region. Eleven people died in a flash flood in August 2020 in Dereli when streams overflowed, destroying many buildings constructed along streambeds and in narrow floodplains.

The word *dereli* means "with streams" in Turkish. And the East Black Sea Region is indeed a region of streams. It hosts a vast network of rivers and streams, small and large, flowing from high mountain ranges to the Black Sea. They have in common their steep slopes, enabling their rapid downhill flow, bursting with kinetic energy. Small-scale, mostly run-of-the-river HEPPs that have been popping up in each and every valley of the region aim to capture this bursting energy in the form of electricity. Eyüp Mumcu, the chairperson of the Union of Chambers of Turkish Engineers and Architects (TMMOB), blames the Dereli disaster on those run-of-the-river HEPPs and small-scale dams, along with the unregulated development of the town, which goes so far as to build on and along the streambeds. The excessive HEPP development destroys river ecosystems and their natural water retention capacity, he says, and deforestation caused by HEPP construction contributes to soil erosion, leading to floods

(Akduman 2020). There were thirty-eight HEPPs in Giresun Province alone, and another seven were under construction at the time of the flood. Along with mining projects, stone quarries, and new highways, HEPPs have been radically altering the region's geography over the last two decades (Yackley 2020).

THE JUSTICE AND DEVELOPMENT PARTY, RENEWABLE ENERGY, AND BULLDOZER CAPITALISM: THE RISE OF HEPPS IN TURKEY

The long, multilayered history of the relationship between hydropower, national and regional development, and modernization has been explored from many different perspectives (see, e.g., Fahim 1981; Ribeiro 1994; Mitchell 2002; Khagram 2004; Billington and Jackson 2006; Kaika 2006; Klingensmith 2007; Sneddon and Fox 2011; Swyngedouw 2015; Tilt 2015). The social and environmental impacts of megadams (see, e.g., Goldsmith and Hildyard 1984, 1986; McCully 2001) have been ignored in a rush to catch up, especially in the so-called Global South, to rise up "above the contemporary level of civilization," as Mustafa Kemal Atatürk put it in 1933 in his speech delivered for the tenth anniversary of the Turkish Republic.[1]

In the last few decades, however, mega hydroelectric dam projects in such developing countries as India, China, Brazil, and Turkey have been subjected to much criticism and challenged by vigorous transnational campaigns. Even though most of these campaigns have ultimately failed to prevent the construction of megadams such as Sardar Sarovar in India, Belo Monte in Brazil, Three Gorges in China, and Ilısu in Turkey, they have been successful in raising global consciousness and, in some cases, blocking international loans and credit guarantee schemes (see, e.g., Eberlein et al. 2010).

While the global movement against megadams was celebrating this relative success in forcing international institutions such as the World Bank to reconsider their unconditional financial support for megadams, the rise of the "renewable energy" discourse gave a new spin to the discussion. The rising challenges of climate change provided developing countries with justification for investing in hydropower. The shift in hydro-power giant China's discourse from economic growth to environmental protection, renewable energy, and sustainable development marks the trend (Lee 2014). As hydropower plants have become an essential com-ponent of low-carbon energy schemes (see, e.g., Baird, Shoemaker, and Manorom 2015), the climate change discourse has given hydropower the "kiss of life," despite its well-documented adverse impacts on ecosystems and communities.

Turkey's move toward rapid hydropower development manifested itself at a historical conjuncture—the first decade of the 2000s—when various narratives of privatization of public assets and private enclosure of com-mons, the energy- and construction-driven growth strategy of Turkey's Justice and Development Party (AKP) government (Erensü 2018),[2] and climate change mitigation intertwined around the issue of hydropower. Privatization of the energy market in Turkey was initiated just before the AKP came to power in 2002. In fact, the establishment of an open energy market, which would be regulated by an independent public institution (the Energy Market Regulatory Authority, EPDK) in Turkey, was a condi-tion set by the International Monetary Fund and the World Bank to release credit after the economic crisis of 1998–99.[3] The Turkish Electricity Market Act, passed in 2001, authorizes the EPDK to grant licenses to private companies to build HEPPs all over the country for elec-tricity production. These licenses allow private companies to sign agree-ments with the State Hydraulic Works (DSI) to purchase the use rights of river waters for forty-nine years. With the passage of the Water Use Agreement Act in 2003, which regulates the terms of such agreements between the DSI and private companies, the legal infrastructure required for private HEPP development was finalized, more or less simultaneously with the establishment of the AKP government.

The AKP's energy- and construction-based growth strategy involves a drastic transformation of urban and rural landscapes, ignoring previously effective protection measures. Energy and construction projects are thus vehicles of "bulldozer capitalism," aptly defined by Erdem Evren (2022, 6–7) as "a system of power, profit and hegemony that comes to be formed and reproduced through the destruction and recomposition of the physi-cal environment."[4] Bulldozer capitalism has also proved conducive to transferring public—and common—resources to pro-AKP firms through excessive use of privatization and public procurement mechanisms. Granting licenses for energy and mining projects to loyal supporters is an essential aspect of the AKP's strategy of "building a loyal business class through an elaborate system of rewards and punishment since 2002" (Esen and Gumuscu 2018, 351).

As small-scale, run-of-the-river hydropower projects do not require much capital or know-how, they emerged as profitable investments for small and medium-sized enterprises that previously operated in labor-intensive sectors such as textiles, providing the enterprises with an opportunity to reconfigure themselves as private energy companies. In addition, Turkey's recognition of all HEPPs and dams with a capacity of less than twenty megawatts as sources of "green energy," following the EU's directive on subsidizing renewable energy sources, led to a series of incentives and support schemes that promoted HEPP construction.[5]

The AKP government has been effectively synthesizing the narrative of energy independence—which equates growth and development with energy production—with a renewable energy brand to promote private run-of-the-river HEPPs and neutralize the strong opposition it faces.[6] As Turkey's president Recep Tayyip Erdoğan put it,

> Before our reign, they used to say, "Water flows, and a Turk just watches it." We have reversed that. Now water flows and a Turk makes [energy out of] it. . . . We need more energy in parallel to our growth. . . . The amount of electricity a country produces shows its level of development—it means that our factories are working, that production is increasing, that our consumption is rising, that technology is becoming widespread.[7]

Besides Erdoğan, other members of the government, such as Veysel Eroğlu, who served as minister of Forestry and Water Affairs for more than a decade from 2007 to 2018,[8] have been influential in framing HEPPs not only as a clean, cheap, and healthy renewable source of energy, but also as the magic solution to Turkey's energy and current account deficits.[9] Consequently, HEPPs have been presented as a major tool to "catch up," ending Turkey's energy dependence, and thus as an issue of "national interest."[10]

Turkey is famously in the grip of an increasingly authoritarian and repressive political regime identified with the AKP government and Recep

Tayyip Erdoğan. While space limitations do not allow me to discuss Turkey's recent past in detail here (see Yılmaz and Turner 2019; Özyürek, Özpınar, and Altındiş 2019; Babacan et al. 2021), it is worth reiterating the intensity of this grip, as Joost Jongerden (2022, 1) does in his introduction to Routledge's recent *Handbook of Contemporary Turkey:*

> In the last few years alone, Turkey has experienced a failed coup attempt; a prolonged state of emergency, only to be lifted after the implementation of a presidential system based on the supreme power of the head of state and the absence of proper checks and balances; a crackdown on traditional and new media, universities and civil society organizations; the detention of journalists, mayors and members of parliament; the establishment of political tutelage over the judiciary; and a staggering economic crisis.

One can imagine that it is not common for members of the AKP government during its twenty—and still counting—years of rule to voice even slightly different opinions regarding HEPPs or any other matter. While some do dare to do so,[11] their moderate positions have never been adopted by Erdoğan, who has steadfastly embraced the issue to the point of accusing the local and environmental opposition of treachery. Declaring himself the "finest environmentalist,"[12] Erdoğan has frequently accused "those environmentalist types" of trying to prevent Turkey's great leap forward in energy, industry, and technology by "disseminating false information and lies" about trees being cut and creeks drying up.[13]

"GOD'S WATER": RELIGIOUS FRAMING OF THE ENVIRONMENTAL COMMONS AS RESISTANCE TO EXTRACTIVE CAPITALISM

Contributing to ecological disasters such as occurred in Dereli and destroying river ecosystems and natural habitats, HEPPs also dispossess human communities that have lived along rivers for centuries. A reminder is in order here: even after decades of rural-to-urban migration within Turkey, and despite the adverse effects of neoliberal agricultural policies that reduce or eliminate public support mechanisms such as state production and distribution facilities, market protections, and fixed-price purchases, agricultural activities, especially smallholder farming in rural areas, continue to be a significant source of income and family provision (Öztürk, Jongerden, and Hilton 2018; Öztürk, Jongerden, and Hilton 2021). It is crucial here that private ownership of agricultural land is not concentrated in the hands of a few (whether companies or landlords), with the possible exception of certain parts of the Southeast Anatolia Region (the predominantly Kurdish region). Smallholder family farming is the norm, with 90 percent of an estimated three million farming units having less than 20 hectares of land (two-thirds have less than five hectares) (Öztürk, Jongerden, and Hilton 2021).

During my ethnographic research, it was common among rural communities across Anatolia that were fighting against HEPPs to frame river waters as "God's water" and "God's river" (*Allahın suyu* and *Allahın deresi* in Turkish), as "God-given" and owned by no one but "God": "Who could take God's water away?" "My beautiful God gave these beauties to us!" "This nature is God-given, and no one but God can take our river, our nature away!" "God created these rivers; God gave them to us." Words such as these came up very often in my conversations with local villagers, mostly women but also men, who resist hydropower projects in and around their villages, which are often quite remote, "hidden" within deep valleys in various parts of Anatolia. They reflect a commonsense expression of their claims that the waters of the rivers cannot be owned or sold. They should, instead, be protected as they are, as the way God created them and "gave" them to the peoples of the riverside areas. Hence, this framing of "God's water" expresses not only a particular belief in a monotheistic religion (in this case, Islam) and a certain image of God as the eternal possessor of all creation—*Malikul-Mulk* or *Maalik Ul-Mulk*, one of the ninety-nine names of Allah—but also a political position regarding the ownership claims of environmental entities on the Earth.

This narrative of flowing water as belonging to God or being God-given is deeply rooted in the oral histories, stories, and myths of Anatolian geography. It is depicted in a famous scene of the film *Susuz Yaz* ("Dry Summer"), a 1964 Turkish drama directed by Metin Erksan that won several awards, including the Golden Bear at the 14th Berlin International Film Festival. It is also used as the opening scene of *Sudaki Suretler* ("Figures in the Water"),[14] one of many documentary films about antihydropower struggles in Turkey that have been produced in the last two decades. The scene shows an argument between the villagers and the landlord Osman, who decides to dam the spring on his property. In the scene, Osman announces to the villagers that he will not be allowing the spring waters to flow toward the village anymore. The villagers react by asking if he understands what he is saying. He responds, "Doesn't the spring belong to me? It is on my land. I have the deed to the land. I can do whatever I want. You have to take care of yourselves." The villagers respond one by one, providing an array of arguments that are still relevant in the struggle against HEPPs: "Water belongs to everyone. How can you claim ownership of water that flows on Earth?" "How can water be yours? God owns it." "This water has been flowing here since the time of our father Adem [the Turkish rendering of the biblical Adam]. No one has the right to claim it for himself." "Water is the blood of the soil. You want to cut our blood off."

I interpret this widespread discourse of the "God-givenness" of the river waters as a religious conception of the commons, which indicates an accurate understanding of the (environmental) commons as "nonproperty" (Hardt and Negri 2017; Mezzadra and Nielson 2019). Even though the vocabulary of the commons is not explicitly employed, the status of water as an environmental commons is a central theme in the antihydropower struggles in Turkey, shared by both local communities and a variety of movements, organizations, and networks. The shared argument is that water—river water in this case—is essential to human and nonhuman life, cannot be turned into a commodity, and, thus, cannot be bought or sold. While environmentally motivated national networks such as *Su Hakkı Kampanyası* ("Right to Water Campaign") have invoked the vocabulary of water as a human right (Su Hakkı Kampanyası 2016), situating, locating the Turkish case within a global legal, political, and discursive framework of water rights (Gleick 1998; Bakker 2007; UN 2010; Sultana and Loftus 2013), more political-economy-oriented groups such as *SuPolitik* ("WaterPolitics") and *Suyun Ticarileşmesine Hayır Platformu* ("No to the Commercialization of Water Platform"), on the other hand, have directed their resistance efforts against the commodification of water (and nature) (STHP 2012),[15] in line with opposition to neoliberal processes of profit-based growth and privatization, thereby locating the issue in the broader anticapitalist and eco-socialist frame (see, e.g., Foster 1999; Foster, Clark, and York 2010; Ricoveri 2013; Klein 2014; Moore 2016).

Beyond the frames of these movements, however, the idea of the commons has proved to have a conceptual and practical capacity to represent popular aspects of the antihydropower struggles, that is, the very motivations of local people to oppose and fight against HEPPs. The commons, as shared spaces, spheres, relations, and resources, do not belong to private persons, nor do they belong to the state (Hardt and Negri 2009; Caffentzis 2010). Hence, they resist the idea of ownership, property, and commodity; they do not belong to anyone and are open and accessible to everyone.[16] This idea of the commons as belonging to no one and to everyone powerfully resonates with the struggles of many local and Indigenous communities to claim and protect their immediate environments: waters, lands, and forests. They might not refer explicitly to the conceptual discussions on and around the notion of the commons, but they possess both the knowledge and the experience of "commoning" (see, e.g., Nightingale 2019). Take Cahit, who, as the elected village head (*muhtar*) of the village of Arılı in the Fındıklı District of Rize Province (EBR), was the highest local administrative authority in the village when I interviewed him in the summer of 2013. It is highly unlikely that he had come across the notion of "the commons," but his definition captured the very essence of the discussion:

> Spaces/places of life [*yaşam alanı*] belong to people who live in them. So if we live in this valley, this nature, this water, this air, this sun is ours. But when we say they are ours, we do not mean that we own them as our property. We don't have the ownership of them; we don't have them as our property. But they are *ours*.

This definition of the environmental commons runs counter not only to the private enclosure of such places, but also to any sense of their "public" ownership. The idea that the state is the legitimate owner of all the rivers in the country and, thus, can transfer the usage rights of river waters over to private persons and companies is rejected by the local communities that have lived with those rivers for generations. Hence, the framing of a river as "God's" represents a discursive resistance to the state's claim of ownership over river waters. God's property is not transferrable; you can only *use* what God gave you; you cannot *sell* it. In such a religious framing of the commons, God has "ownership" of nature as God "created" the Earth in the first place. What, then, is the justification for the state's claims of ownership?

In a country such as Turkey, where the infrastructural power (Mann 1984) and social capacities of the state are limited and where the accent is on controlling the territory and the people more than providing for them (see, e.g., White and Herzog 2016), the state is like a despotic father who has not done anything to raise a child properly but feels entitled to exercise unrestricted authority over that child's life. It is thus a strange experience that the state, which does not come when needed (i.e., it does not provide for the real infrastructural needs such as transportation, education, health care, etc. in the remote, geographically secluded rural areas of the country), now suddenly claims the river, which it has only seen on maps, just to sell its waters to private companies. "As if the state ever gave us anything! And now it wants to take our river away from us," says Nuran, a woman in her seventies from the village of Aslandere (EBR).[17] In this context it is crucial that the expressions "God's river" and "our river" are used interchangeably. It expresses that what is God's is also ours (like the commons); what is the state's, on the other hand, is not ours, as the current wave of privatization of public spaces, resources, and assets has demonstrated.

Privatization is only one aspect of a broader process of the expansion of capital (as a social relationship) and nature into previously nonmarketized social and ecological spheres, spaces, relations, and resources—into the very "webs of life" (Moore 2016). It is not a coincidence that Polanyi has become popular again in recent decades, especially after the 2007–2008 financial crisis, as his concept of fictitious commodification—of nature, labor, and money—points to the colonizing tendencies of unregulated markets in the early twentieth century (Polanyi [1959] 2014). Contemporary scholars such as David Harvey (2003), Klaus Dörre (2012), and Nancy Fraser (2014) discuss why and how the age-old concepts of Marx, Luxemburg, and Polanyi—the concepts of primitive accumulation, *Landnahme*, and fictitious commodification, respectively—can still be used to analyze the present-day conjuncture of capital expansion and enclosures of the commons. An important outcome of these discussions has been the conception of expropriation not as an exception used for institutionalizing private property in a specific spatiotemporal context, but as an essential mode of capitalist accumulation that goes hand in hand with "normal" mechanisms of valorization (Fraser 2016; Mezzadra and Nielson 2019).

This idea of primitive accumulation as an ongoing process, primarily through Harvey's concept of "accumulation by dispossession,"[18] has become popular among scholars who work on environmental struggles (or, as political ecologists prefer to put it, resource conflicts), as it concisely depicts what happens on the ground, that is, how local communities are dispossessed—of lands, waters, forests, and clean and healthy environments—through various processes of expropriation. Expropriation of nature provides private firms with natural "assets," that is, "nature's free gifts" (Moore 2016; see also Foster and Clark 2018) to enhance new productive capacities at very low cost, as in the example of small-scale hydropower, whereby a company uses river waters to produce electricity. This inclusion of nature's "free gifts" into the commodity system through the plundering of previously noncommodified ecosystems, entities, relations, and the spaces that we call "nature" is a fundamental form of the market's expansion in our time and a central aspect of "extractive capitalism" (see, e.g., Petras 2013; Webber 2015; Mezzadra and Neilson 2019).[19]

Finance acts as a form of extraction, as financial instruments deepen the process of commodification, layering it with and diffusing it among various institutional structures, spaces, and temporalities. In the case of water, since HEPPs are recognized as principal sources of clean, renewable energy, they have become a popular element of carbon offset schemes, along with energy efficiency, forestry, and waste diversion projects. By investing in carbon offset projects, private companies obtain carbon credits, which are financial instruments that provide the buyer with a right to emit more carbon dioxide into the atmosphere. Their investments in carbon offset schemes allegedly compensate for the carbon emissions released elsewhere in the world.[20] If we put it in Polanyian terms, the fictitious commodification of money, and its contemporary form as financial instruments, is the key to the fictitious commodification of nature (land) and social production (labor) in our age. This finance-assisted commodification of nature is very much state sponsored: states act as legal and institutional facilitators to the corporate powers of capital, as discussed below in the context of antihydropower struggles in Turkey.

UNIMAGINED COMMUNITIES: DISPOSSESSION, DESTRUCTION, AND DISCOUNTING

The first few years of HEPP development revealed the socio-ecological destruction they bring about in many different localities. One main problem was the extensive and unregulated nature of HEPP construction. It was common to license multiple HEPPs in one river basin, sometimes more than ten, as in the İkizdere and Senoz Valleys in the East Black Sea Region, the execution of which would mean that river waters would be transferred from one pipeline to another, barely touching the soil. Licenses were typically granted on paper in Ankara, without an adequate social and environmental impact assessment, even without sufficient geographical knowledge of the region in question. Around 2010, the number of licenses and constructions peaked, with a target of 4,000 HEPPs to be reached by 2023 (Turkish Water Assembly 2011; Gibbons and Moore 2011).

Ineffective environmental assessment and monitoring, along with poor project design and implementation, have augmented the destructive effects of HEPPs (Kömürcü and Akpınar 2010; Şekercioğlu et al. 2011), accelerating habitat degradation caused by tree cutting, excavation and dumping of the construction material, fill areas, blasting, construction of roads, pipes, and water storage systems (Başkaya, Başkaya, and Sari 2011). In the end, remote Anatolian valleys were being turn one by one into construction sites. Bulldozers, excavators, and dump trucks were everywhere, digging up the earth to build roads and install pipelines and turbines, and leaving the excavated soil and rocks in the riverbeds or moving them over high hills and dumping them down into the valleys, causing soil erosion (see Kurdoğlu 2016 for a detailed discussion and pictures).

The areas that are degraded by HEPP construction are, however, not deserted. They have been inhabited, often for centuries, precisely due to their location in river basins. Adding insult to injury, over and above the loss of river waters the villages located on the riverbanks are being left with devastated forests, radically altered landscapes, dust, dirt, and erosion caused by construction and dumping. Dispossession, thus, goes hand in hand with discounting, as if there are no people living in those riverside villages. Rob Nixon uses the term *unimagined communities* to describe such processes of discounting: the "slow violence" induced by various "resource development" projects, "whether of water, oil and gas, minerals, or forests," which entail "not just the physical displacement of local communities, but their imaginative displacement as well" (Nixon 2011, 150–1).

Such acts of discounting and "unimagining" local communities entail a cultivated ignorance of local communities' everyday lives and practices. Take Turkey's ex–prime minister, current president, and all-time "Chief" Recep Tayyip Erdoğan's famous use of the proverb "Water flows, and a Turk just watches it" (*Su akar, Türk bakar*).[21] Erdoğan often invoked this saying to point out the change brought about by the AKP government. "Now the rivers do not flow in vain," he would say, "as we use and utilize them as electricity-generating resources." Fatma, a young woman I interviewed on her terrace in the village of Pınar in the Yuvarlakçay Valley of the Mediterranean Region, explains how offensive this saying is to them: "As if our use of river waters does not count," she says. "As if we are not living here." It is not only human beings and communities whose use of water is ignored; the decimation of flora and fauna that the river waters sustain is mentioned by many villagers I have interviewed. The same theme appears also in news reports and documentaries made on the issue. In a short documentary film called *Anadolu'nun İsyanı* ("The Rebellion of Anatolia"), a male resident of Senoz Valley (EBR) mentions then–prime minister Recep Tayyip Erdoğan, saying, "Our prime minister says they sold the usage rights of the river waters. Don't I have usage rights? Don't the bear, the fish, all the animals have it? Whose rights are they selling to whom exactly?"[22]

It is often the case that local communities are not even informed about, let alone included in, the processes of deliberation and decision-making about "resource development" projects—such as HEPPs, coal plants, stone quarries, and mines—and their socio-ecological impacts. The overwhelming majority of people I encountered during my field research had not been informed about the planned projects and their prospects through formal channels. Either they had heard rumors that they did not believe—

due to the firm conviction that streams are God-given and belong to the place and the people and thus cannot be enclosed—or they figured out what was happening in the most bitter of ways, by experiencing the construction once it had started.

Instead of informing the village people publicly, state-backed private company representatives often approach those they perceive as influential in the village, usually the *muhtar* (the elected head of the village) and the villagers close to him, offering them various personal incentives and communal benefits to back the project.[23] If and when companies are successful in co-opting such figures in a village and buying their consent, the expression of public opposition becomes difficult, and success becomes less likely due to the division of people into supporters and opponents. This division has had destructive effects on village communities in many cases, sometimes setting members of the same family against each other. Hence, the primary vehicle for gaining consent becomes bribing and dividing the village community, and not public consultation, deliberation, and involvement, which should be an integral part of the process according to environmental impact assessment (EIA) regulations.[24]

The EIA system was adopted by Turkey to comply with the regulations of international law and the EU acquis (Güneş 2020), only to be disabled and reduced to a procedure on paper under AKP rule (Kurdoğlu 2016) to save private investors the headache of regulatory compliance. One crucial strategy in this regard is to reduce the number of projects that require an EIA by legally defining more and more exceptions to the rule. Constantly amending the constitution, laws, and regulations has been one of the main strategies of the AKP government to establish and maintain its authoritarian rule over the past decade (Özbudun 2015; Kaygusuz 2018; Yılmaz 2020; Petersen and Yanaşmayan 2020). Environmental policies are no exception. Environmental laws and regulations have been constantly amended to further shrink the scope of nature conservation and open protected areas to investment. EIA regulation itself, adopted in 1993, has been amended countless times to exempt more projects from EIA requirements. Such legal arrangements have provided many projects with a free pass to do as they wish, paralyzing the one and only mechanism through which local communities could legally intervene in construction projects.

Wherever an EIA is inescapable, the Ministry of Environment and Urbanization famously accepts perfunctory EIA reports, often produced without the necessary ground truthing through fieldwork and data collection, sometimes copied and pasted from other reports. Public meetings, which should be an integral part of EIA processes, are intentionally obscured, announced in newspapers no one reads or by hanging a document somewhere in the town center and making sure that no one from the village sees it. There are, of course, very effective ways to disseminate important information in a village, such as loudspeaker announcements, which are often used in villages for official and communal purposes (state announcements, obituary notices, etc.). It is thus telling when a middle-aged peasant woman from Yeşilırmak Valley (Central Black Sea Region) cries out in the documentary film *Sudaki Suretler* ("Figures in the Water"). She is dressed in her everyday clothing—a burgundy headscarf, a blue-gray blouse, and dark-colored *şalvar* (shalwar trousers)—and speaks in a local accent and full of confidence in front of a massive pile of onion sacks, placing one hand on the sacks and gesturing with the other:

> Did they get my consent?[25] Did they announce it over the loudspeakers? I haven't heard anything. How did they appear like that? Even when building a house, you go to the land registry; you bring people to your land; you get approval. What did they do? Did they go from house to house? I have never heard [anything about it]. How could they get [to the river] from here? They could not. I promise God [*vallaha da billaha da*] I will set their cars on fire! If they come again, we will burn their vehicles again. Let them come.

IDENTIFYING REGIONAL DIFFERENCES: THE USES OF WATER AND THE DYNAMICS OF THE STRUGGLE

Even though common motivations, narratives, and patterns are observable within many local communities that fight against HEPPs across the Anatolian peninsula, there are also significant regional differences. The specific dynamics of the (local) anti-HEPP struggle, the framing of the issue and the discourses and narratives employed by the local communities are shaped by multiple determinants. Among them are the ethnic, cultural, and political composition of the local community and the social resources and networks available to them; the specific organization of social and spatial relations; the use of river waters; the material properties of river waters; and the particular characteristics of the connection between river waters, human communities, and the wider nonhuman environment, which goes beyond the river's immediate economic use-value. As those determinants are complex and multilayered, their specific configuration differs from case to case, even from one village to another within the same region. Still, geography proves to be a useful analytical tool for identifying the main differences among the three regions in which HEPP projects and anti-HEPP movements are concentrated: the Mediterranean Region, the East and Southeast Anatolia Regions (which include the Kurdish parts of the country), and the East Black Sea Region.

In the Mediterranean Region, where summers are long, hot, and dry, river waters are essential to subsistence agriculture and animal husbandry. There is a reason why villages are located along rivers across Anatolia. In the Mediterranean Region in particular, river waters are traditionally used to irrigate orchards, vegetable fields, and greenhouses—fruits and vegetables being the primary agricultural products of the region—and the grazing lands that produce fodder for livestock. For that reason, the main motivation behind the struggle against HEPPs is the protection of livelihoods. It is clear that for the villagers in the Alakır and Yuvarlakçay Valleys of the Mediterranean Region, losing river waters would undermine their ability to make a living, most likely forcing them to migrate to urban areas where they would have to join the workforce as unskilled laborers. It is also clear to them that, in a country like Turkey, where unemployment is very high and wages are very low, this would mean impoverishment.

Similar to the time when concepts such as "primitive accumulation" (Marx) and *Landnahme* (Rosa Luxemburg) were first coined to analyze the private enclosure of the commons that resulted in mass proletarianization of rural populations, peasant communities of the rural Mediterranean Region today are threatened with dispossession, displacement, and, thus, forced proletarianization. This aspect of displacement via dispossession is manifested in a conversation I had with Esin, a young woman who lives a very modest life in the village of Karacaören as a smallholder farmer with her father. Even though the Alakır Valley is a natural conservation area, eight small-scale hydropower plants were planned on the river, with a capacity of between two and twelve megawatts each. Four of them had already been constructed by 2014 when I visited the valley. Esin was very active in the struggle to prevent the construction of the remaining four: the Alakır 1 and 2, Çayağzı, and Dereköy HEPPs.

It was clear that Esin and her neighbors would lose the river waters that they use to irrigate their relatively small fruit and vegetable gardens if the remaining projects were not blocked. Esin is a university graduate, but like many others in Turkey, she could not find a job in her field and returned to her village to live with her father. Both their lives and their livelihoods depended on the small amount of agricultural products they raised, as they consumed their products and sold the rest. As a prominent figure in the local anti-HEPP movement, Esin was a part of the committee of local activists and villagers that would regularly travel to Ankara to lobby. They met representatives from the Ministry of Environment and Urban Planning to share their grievances and to present their reasons for opposing the HEPP projects. Esin remembered the bitter response of the undersecretary of the ministry when she asked how they—the peasants of the valley—should earn a living if the HEPP projects took the river waters away from them. He looked at her with empty eyes and said they could work in tourism.

Unlike the situation in the Mediterranean Region, in the East and Southeast Anatolia Regions of the country, especially in the parts where the majority of the population is Kurdish, the entire issue is embedded in the Kurdish struggle for cultural and political autonomy. Engin, a young man from the Mesopotamian Ecology Movement with whom I conducted an in-depth interview in November 2014 in Diyarbakır (the main city in Turkey's Kurdish region, known as Amed in Kurdish), explained his interest in ecological issues, including HEPP projects, drawing on the (presumed) etymological relation between ecology and autonomy with reference to the Greek term *oikos*. Engin is a civil engineer (like myself, an alumnus of the Middle East Technical University in Ankara) who also has a graduate degree in anthropology and visited the Chiapas region of Mexico before settling back in the Kurdish region, where he is originally from. For Engin and other activists in the movement, ecological struggles are an essential dimension of the Kurdish struggle for autonomy. The movement aims to extend the conception of autonomy beyond self-determination, self-sufficiency, and sovereignty over natural resources to articulate the Kurdish people's right and capacity to arrange their own relationship with nonhuman environments.

Regarding the HEPP projects, Engin argued that domination over water is an aspect of cultural and political domination. Most controversial hydropower projects in the Kurdish region differ from the projects in the Mediterranean and the Black Sea Regions as they are not small-scale, run-of-the-river projects, but larger-scale hydroelectic dams. The most well known is the Ilısu Dam, which has been the focus of national and international antidam campaigns, mainly to save the historical city of Hasankeyf (Eberlein et al. 2010; Hommes, Boelens, and Maat 2016).[26] The Kurdish movement became closely involved in the controversy by organizing international campaigns and practically halting the construction through armed assaults and the kidnapping activities of the PKK (Kurdistan Workers' Party). PKK leader Abdullah Öcalan's famous statement equating the hydro dams on Kurdish territory to the atomic bombs dropped on Hiroshima and Nagasaki clearly expresses the Kurdish movement's position.[27]

Besides the Ilısu Dam, the HEPP projects in Tunceli (Dersim) in the East Anatolia Region, located in valleys between high mountains, are also highly controversial. Probably because of its relatively isolated geographical location, Dersim is unique in terms of its ethnic, political, and cultural composition. The history of Dersim is a history of uprisings and massacres, and the place is still very much associated with a leftist-oppositional political culture (see, e.g., Göner 2017; Sözen 2019). The HEPP projects are related to this long history of state violence and oppression and are understood as a tool to redesign Dersim's geography. It is widely believed in Dersim that dam reservoirs would be used to cut off the connections between valleys and, ultimately, depopulate Dersim further (Hamsici 2010).

Even more importantly, the peculiar belief system of the Zaza Alevi population of Dersim entails naturalistic-pagan characteristics, and water, especially the waters of the Munzur River, has a central place in this belief system (Deniz 2012, 2016). It is thus not difficult to understand why the hydro dam projects that threaten the Munzur Valley and the sacred sites located along the Munzur River are vehemently opposed by the local community. The anti-HEPP demonstrations organized in Dersim, especially those in 2009 and 2012, were probably the most well-attended local demonstrations in the whole country, with thousands of participants in a city with a population of around thirty thousand. People were united against HEPP projects, and culture and belief were the central motivations behind the antihydropower struggle. As Ali Rıza, a representative of the Dersim Volunteers for Nature Conversation, told me in 2014 in Dersim:

> Our beliefs are integrated with nature here in the Dersim territory. Our sacred sites are in nature, and sometimes the natural entities, such as the waters of Munzur River, are sacred themselves. And people believe that dams will destroy our holy sites. People believe the submersion of those holy sites will be an insult to our beliefs and an injury to our pride.

While economic arguments around the issues of livelihood and subsistence come to the fore within the anti-HEPP discourse in the Mediterranean Region, political and cultural framings, not surprisingly, are more dominant in the Kurdish regions. However, the East Black Sea Region differs from both, as neither the immediate economic use of river waters nor issues of political autonomy, ethnic identity, and belief drive the anti-HEPP struggles in the region.[28]

The East Black Sea Region is located in the northeastern corner of Turkey, stretching from the city of Trabzon to the Georgian border. It is a region where HEPP development is concentrated and where resistance to those projects is strongest. Even though the East Black Sea Region is populated by Laz and Hemshin peoples who have managed to protect their languages to a certain extent (they speak a Kartvelian language and the ancient Armenian dialect of Homshetsi, respectively), the themes of autonomy, sovereignty, and ethnic and group identity are not employed within the narratives and discourses of villagers or local activists in the way they are in the Kurdish region. This is because Laz and Hemshin peoples are already thoroughly assimilated into Turkish identity politically.[29] Consequently, the region is a stronghold of the governing Justice and Development Party (AKP) and the extreme (Turkish) nationalist party, the Nationalist Action Party (MHP), both organizationally and in terms of the distribution of votes. In addition, even though religion and belief are always somewhat involved in people's conceptions and uses of water (see the notion of "God's water" above), water is not particularly loaded with religious connotations or understood as sacred as it is in Dersim in the Kurdish region, with a few exceptions.[30]

Economic and cultural notions of justice are implied by certain arguments, discourses, and practices of local movements in the region, but they do not predominate. For instance, some villagers voice concerns that the loss of river waters could cause changes in the microclimate, which might affect tea plants, their primary agricultural product. But the economic notion of justice is in no way as prevalent in the EBR as it is in the Mediterranean Region, where river waters are a primary source of irrigation, which is essential for subsistence agriculture. Contrary to the Mediterranean case, livelihoods do not depend on the immediate economic use of river waters in the coastal parts of the East Black Sea Region; rainwater suffices to sustain monocultural tea and hazelnut agriculture. The coastal parts of the region, that is, the northern skirts of the Kaçkar Mountains, are precisely where the HEPP projects, and resistance against them, are concentrated.[31] Hence, it would be misleading, at the very least, to portray the anti-HEPP struggles as "resource" conflicts driven by livelihood concerns and to explain them away as being caused by economic and distributive notions of justice.

What, then, drives the strong resistance against HEPPs in the East Black Sea Region? The coming chapters aim to develop a response to this critical question, building on the empirical material I have presented and started to analyze in this chapter. However, to contextualize the detailed discussions in the following chapters, it will be worthwhile first to scrutinize the organization and conduct of local anti-HEPP struggles.

ORGANIZATION OF RESISTANCE: LEGAL
BATTLE AND SOCIAL STRUGGLE

As soon as local movements started to organize themselves against HEPP projects affecting their villages and valleys, notably around 2007–8, many supportive networks and organizations emerged to support their struggles, such as Su Meclisi (Water Assembly), Su Hakki Kampanyasi (Water Rights Campaign), Suyun Ticarileşmesine Hayır Platformu (No to the

Commercialization of Water Platform), and Karadeniz İsyandadir (KIP, Black Sea in Resurrection). While these networks served as public platforms bringing academics, lawyers, and national and regional activists together, affected communities established their own organizations at the local level. The spatial scale of local organizations varies from a single village—such as the Boğazpınar Village Platform against HEPPs (Mediterranean Region, MR)—to a single river threatened by multiple projects—such as Let Munzur Flow Free Assembly (East Anatolia Region, EAR)—to a regional center where multiple villages or valley communities fight against multiple HEPP projects together—such as in the case of the Fındıklı Brotherhood of the Rivers Platform (EBR).

The specific configuration of relations between these networks, mainstream environmental organizations, and local communities on the ground varies significantly from case to case. In general, these organizations tend to identify with one of two fundamental lines of argumentation: against ecological destruction or against the commodification of water. Even though these arguments are often mutually reinforcing, clashes between organizations that find themselves emphasizing different approaches—either environmentalism or anticapitalism—are common. As a result, the constellations of networks established by local organizations can vary significantly from one place to another. Some of them receive support from mainstream organizations such as Doğa Derneği (Turkish Nature Association) and TEMA (Turkish Foundation for Combating Soil Erosion). Others are wary of such support and dissociate their struggles from mainstream environmentalism due to mainstream environmental organizations' ties to private companies, some of which they claim are involved in HEPP projects (see Eryılmaz 2018).

The voluntary labor of local activists—that is, local people with certain social and cultural capital who commit their time and energy to the anti-HEPP cause—is crucial to establishing and maintaining local organizations. These local activists are overwhelmingly male, primarily left-leaning, educated members of the local communities. They play an instrumental role in establishing crucial links between the networks mentioned above and local people through the local organizations they help to build. In the last ten to fifteen years, such organizations have put their imprimaturs on thousands of local meetings, panel discussions, press announcements, and demonstrations, intending to inform and organize affected communities against HEPP projects. The anti-HEPP movement represents perhaps the first time in the history of modern Turkey that so many academics and professionals (environmental and forest engineers, ecologists, economists, sociologists, and geographers) have traveled to so many villages in rural Anatolia to inform villagers about an issue. Although most of the peasants living in those villages were attending panel discussions, meetings, and demonstrations for the first time, many actually outperformed the professional activists as they exchanged views with university professors, environmentalists, and lawyers on an equal footing, expressed their opinions to journalists and researchers, and physically confronted private companies and state forces.

These local organizations cooperated to differing degrees, and some of them joined the East Black Sea–centered national network Brotherhood of the Rivers Platform. Local communities and their representatives came together at festivals, meetings, and demonstrations at both the regional and national levels. They exchanged information and learned from one another, especially regarding various forms of struggle and movement tactics, both in these encounters and through widely circulating social media posts. For example, the tactic of camping out at the site of a planned project day and night to prevent passage of construction vehicles has been widely used since it was "invented" by the Pınar and Beyobası villagers, both located in the Yuvarlakçay Valley (MR).

Berna Babaoğlu Ulutaş, a young woman who has served as the pro bono lawyer for the Pınar and Beyobası villagers, told me the interesting story of that "invention" on a sunny autumn day in 2015 as we were talking in a seaside café in Fethiye, which happens to be my hometown. The story started when the HEPP company cut down monumental trees on the banks of the Yuvarlakçay River, which were supposed to be protected by law, to make space for the construction. Berna remembered it vividly: "When we went there [with the villagers] afterwards, the trees were still there, lying around like corpses. It affected us all." Those trees, the enormous stumps of which I visited before talking to Berna, were ancient (around three hundred years old), and people had a particular connection with them. It was a special place of rest and worship, a place where people would go to have a picnic but also to sacrifice an animal or to pray. Berna continues:

Women sat on the trunks [of trees that had been cut down]. "We will not give these trees to the company," they said. "These are our corpses, this is our funeral, and we won't let them clean up the space to start construction." They needed to keep guard, then. Day and night. But how? It gets cold at night. So, bringing tents became a necessity. Some people slept in their cars the first nights. Then they started to build stuff: a kitchen, a toilet, storage for food. People began to live there.

In the documentary film *İşte Boyle* ("It's Like This"; English title: *Damn the Dams*), people from Bağbaşı (EAR) tell a similar story of being forced to establish a protest camp because the construction workers developed the habit of working after midnight when the villagers were asleep. As a result, people started to keep guard at the construction site twenty-four hours a day, and that is how a camp started to emerge.[32] As a form of protest, the protest camp was invented out of necessity, in line with the requirements and practicalities of everyday struggle (see Yaka and Karakayali 2017). They then became a symbol, a trademark of the anti-HEPP movement, a place (or, more accurately, multiple places in multiple regions of the country) where many NGO representatives, local and national politicians (from opposition parties, of course), and people from all over the country came to show their support.

As in the case of Berna Babaoğlu Ulutaş, most of the local communities that oppose HEPP projects have had the pro bono legal support of environmentalist lawyers within arm's reach. The legal support of such lawyers, who established the independent network CEHAV (Lawyers of Environment and Ecology Movements) to share knowledge, information, and legal strategies, has been crucial in legally blocking or suspending many HEPP projects. Most of them start their working day by checking the website of the Ministry of Environment and Urbanization's Directorate General of EIA Permits and Inspection, despite their busy schedules. They steal time from cases that earn them money to devote their time and energy to pro bono work. Thanks to their voluntary work, efforts to disable the EIA procedures can still be resisted and, thus, the EIA stipulations can still be used as leverage for local communities to legally stop or suspend the HEPP projects by objecting either to the perfunctory EIA reports or to the exemptions granted by the ministry or the governorates. However, even the lawyers themselves agree that what determines the movement's overall success is not the legal processes, but the social struggle.[33] They see their work as a part of local communities' struggles, not as a substitute for them.

Most of the court decisions that suspend HEPP projects were based on expert reports that challenge the EIA approval reports by demonstrating the prospective social and environmental impacts the projects would cause. It has been proven, though, over and over, that strong local opposition affects court decisions, probably because local movements put pressure on courts by demonstrating those negative impacts in practice. It has also been the case that some companies withdraw even before a court decision is issued due to the strong local opposition, as happened in Yuvarlakçay. Local opposition also acts to safeguard and enforce the legal success, as many companies, relying on the strong support of the AKP government, tend to continue construction work despite court orders to cease (as happened in the Senoz Valley, EBR).

In fact, government support has been so unconditional that Erdoğan famously opened the Cevizlik HEPP, located in the İkizdere Valley (EBR), even as the legal process regarding the project was still ongoing. He turned it into a spectacle of attacking environmentalists and protesting communities, accusing them of disseminating false information and outright lying (Avcı and Kaçar 2010). The AKP's attitudes regarding HEPP conflicts have contributed to its widespread image as a government that ignores legal decisions when they do not serve its purposes and otherwise intervenes in legal processes, both by constant legislative and constitutional amendments that progressively erode judicial independence and by switching judges during lawsuits (as routinely happens in HEPP cases).

This does not mean that legal processes are of no importance, however. They are important insofar as they arm the resisting communities with legal legitimacy, which bolsters the moral and social legitimacy these communities have already achieved by effectively communicating their cause in the public sphere. As Ahmet, a local entrepreneur and a well-known activist against HEPP projects in the Alakır Valley (MR), told me on a hot summer day in Kumluca town center in 2014, "You need to have [legal] papers in your hand." By *papers*, he meant documents that verify juridical controversies or victories, providing the local communities with lawful authority as they resist HEPPs. Still, in a political environment that openly flaunts the rule of law, "papers" play only a supporting role. It is the insistent and durable local movements that can safeguard legal decisions by making it clear to the HEPP companies that things will be difficult for them if they do not obey the court orders.

Local communities have learned this lesson well, not only through their struggles but also from the experiences of other communities. "Courts make a decision, but it is we who determine the real result," proclaim the people of Aksu Valley (EAR) (Hamsici 2010, 227). Bahattin Nacit, the *muhtar* of Çamlıkaya village in Aksu Valley, explains how they came to such a conclusion:

> We are following all the developments regarding HEPPs via the Internet and media. What we notice in other places is that it seems a social movement is more effective than legal processes. Frankly, the legal procedures seem to have degenerated. Even if a court decision suspends a project, the companies continue working in many places. For example, in Muğla [Yuvarlakçay], they cut 200–300 ancient trees in one night, even though they were under protection. And people started to keep guard at the construction site after that. So that's how you do it. (Hamsici 2010, 229–30)

CEHAV founder Fevzi Özlüer stresses a similar point using different language:

> The only way to force the administrators and the capitalists to act in compliance with the principles of the rule of law in countries like Turkey is the self-power of communities. . . . Court rulings open the space for a public discussion. Legal processes are like antidepressants for politics; they repress the problem and allow you to think and act. If you cannot use the political tools at this time, going back to the pill does not solve the problem when it explodes again. (Özlüer quoted in Erensü 2016a, 466)

And, indeed, the problem explodes again and again. Some companies just start the entire EIA process all over again after their first EIA approval is canceled by a court ruling. Or licenses change hands from one company to another. As a result, one never knows when a given HEPP project will reemerge from its ashes like a menacing phoenix. Thus, one can never rely on one's initial success. In many localities where the initial attempts to construct a HEPP had been successfully blocked, such as in Loç Valley (WBR), Boğazpınar (MR), and Fındıklı (EBR), local movements had to fight against constant attempts, by either the same or different energy companies, to reinitiate the construction process, despite court rulings and widespread protests.

In some cases, such as in Loç Valley, local community members had to stand trial on charges such as insulting company representatives and blocking construction work. In the meantime, the Council of State (*Danıştay*) twice overturned the local court's decision to suspend the HEPP project, and the local court issued the same decision a third time. As they were waiting for the Council of State's decision yet another time, Halime Çakmak, a middle-aged woman and one of the eighty-four members of the local community who faced trial, made a statement that affirms Fevzi Özlüer's point on the socialization of legal processes: "We are all more informed now. Most of us have become like lawyers" (Özlüer quoted in Köse 2016).

A tactical legal battle continues between the local movements and their supporters and the government-industry coalition. While the local movements and their lawyers keep inventing new ways to obstruct the projects using the tools available in domestic and international law, the government–industry coalition responds by making those tools inoperable by changing the law, switching out judges, or otherwise complicating the legal processes. Manipulating the legislative powers of the government to reflect the requirements of the newly flourishing HEPP industry appears to be an effective tactic used by the government-industry coalition to disarm the movement on the juridical level. The (ironically named) Draft Law on the Conservation of Nature and Biodiversity was a juridical ploy to open conservation areas to extractivist industries.[34] Such grand measures as drafting new laws go hand in hand with routine uses of existing legal tools, such as eminent domain to expropriate land that belongs to villagers, and increasing legal expenses, such that filing a lawsuit becomes not only more complex, but more expensive as well (Erensü 2016b).

There is also another aspect of the battle between these camps that takes place on the ground. It involves physical attacks and clashes with the police, the gendarmerie (armed state forces in the villages), and the security personnel of the private companies. As mentioned before, many peasant communities around Anatolia organized and attended demonstrations, sit-ins, or protest camps for the first time in the context of the anti-HEPP struggles. As a result, many of them faced legal charges, not to mention violent consequences such as injury by tear gas, batons, and plastic bullets. Such a violent encounter cost Metin Lokumcu, a middle-aged man and a retired teacher, his life. Lokumcu suffered a stroke due to overexposure to tear gas (as confirmed by the Turkish Medical Association) during an anti-AKP and anti-HEPP demonstration in Hopa (EBR) on May 31, 2011. Another victim, Ahmet Türkkan, was in his eighties when he died of a heart attack after being taken by the gendarmerie to the station, where he was kept waiting for hours until his statement was finally taken. Along with other villagers, he was trying to prevent employees of the HEPP company from taking measurements for construction work in the village of Karacaören (MR). He returned home from the gendarmerie station angry and tired and died the same day, August 17, 2010.

Such violent encounters and incidents have had traumatic impacts on those communities and provoked political and emotional responses that vary from disappointment to alienation. As many locals I spoke to repeatedly emphasized, it is about discovering that the state belongs not to you, but to the private companies. The people who are supposed to protect you attack you instead; they act as if they are the armed forces of the private companies, behaving no differently from those companies' private security personnel in most cases.[35] And you suddenly find yourself on the other side, being framed as a criminal, a looter, a traitor—as the enemy.

Various journalistic accounts report the dramatic effects of such experiences. Mahmut Hamsici, for instance, met the young daughters of Sezgin Yılmaz. Yılmaz was seriously injured by a HEPP company's workers and security personnel in the village of Düzköy (EBR). Yılmaz and his friends wanted to block the construction work, as the legal process regarding the HEPP project was still pending. Even though he was severely beaten and had to be hospitalized, the attackers faced no consequences; the gendarmerie unit that arrived after the incident took no measures. The same unit, however, proved very quick to the scene when Hamsici was conducting interviews with villagers. They asked whether he had permission to conduct research. When they were gone, the children asked: "Who were they?" When they learned that they were soldiers, they asked: "Soldiers of the company?" (Hamsici 2010: 67–72).

Finding themselves in such a novel constellation, people suddenly start to identify less with the majoritarian state and more with minorities whose existence is systematically denied and violated. Parallels with Native Americans and Palestinians are voiced; even the politically taboo subject of the Kurdish struggle against the Turkish state became a point of reference, as in the case of young men in Karacaören village, located in the Alakır Valley (MR), who told me that people were forced to take arms and take to the hills like the PKK.[36] It was not the case that those young men were sympathetic to the PKK; on the contrary, they were primarily Turkish nationalists.[37] Rather, the analogy to the PKK indicated the scope of the disillusionment and alienation from the Turkish government they had been experiencing.

The cause of alienation could be the state—as Seniye, a young woman from the village of Aslandere in Fındıklı District (EBR) who also defines herself as a Turkish nationalist and a supporter of the ultranationalist Nationalist Action Party (MHP), concisely expressed: "If the state is not with us, then we are against the state as well." This could just as well refer to the AKP, a party strongly supported in the rural areas of the country. At the beginning of the documentary film *İste Böyle* (*Damn the Dams*), a group of peasant women from Tortum (EAR) explain how they were beaten up and dragged by the police during their protest at the HEPP construction site and how some of them passed out. The film also graphically depicts this happening with archival footage filmed during the protest. Near the end of the documentary, an older woman talks about how her admiration for Erdoğan has turned into disappointment, referring to the HEPP project and the police attack they suffered. Her words illustrate the political effect the conflict has had on certain parts of the rural population that resist HEPP projects. She says:

> He seemed to be good. We voted for him. Look what he did to us. He takes the water from the fields, the trees, and even the drinking water. How are we going to live? . . . He had us dragged up there in the mountains. I do not give my blessing to him. Neither in this world nor in the afterworld. Let him be hanged upside down in hell! . . . I used to pray for him day and night. Now I curse him! . . . He will dry us out. We made him our leader. We had it coming to us.

The effect of this disappointment and alienation is by no means universal, though, as many members of those communities continue to support the

AKP and Erdoğan, believing it is not Erdoğan who imposed HEPPs on them, but the private companies and some state officers working behind Erdoğan's back. In some cases, however, such as in Fındıklı, one of the symbolic places of the anti-HEPP resistance, the struggle has resulted in a change in local government, as people reacted against the AKP by voting for opposition candidates who supported their cause.

2 Resources, Livelihoods, Lifeworld

LINKING GENDER AND ENVIRONMENT THROUGH THE LIVED BODY

We won't give up on our river. We won't! We won't! We
won't! There will be blood. I will sacrifice my life if it comes
to that. If I jump in front of a bulldozer and die here, these
villagers will not sit with their hands tied. . . . To the death,
my daughter, [we will fight] to the death!

Semra, a middle-aged woman from the village of Gürsu

It is I whom Semra calls "my daughter." I recorded her words above as we
were talking in front of her village house by the Arılı River in the village of
Gürsu in the summer of 2013. Semra's passion was not unusual. I encountered many more who talked like her. One cannot help but be struck by
women's resilience and courage in all the local, grassroots struggles against
HEPPs (as well as against coal plants, gold mines, stone quarries, and
many extractivist projects all over Anatolia), as it is the women who talk
about risking their lives and even killing for the cause, while most men try
to adopt more "reasonable" language. Using confrontational language,
women openly challenge the legitimacy not only of the private companies,
but also of the state apparatus to enclose and transform their immediate
environments. An elderly woman from the district of Çamlıhemşin (EBR)
became famous as Mother Havva (*Havva* is the Turkish rendering of
"Eve") when she said, "The governor says that we are looters. Who is he? I
am the people, and I am here. Take your machines and go away." She was
sitting alone in front of a military garrison and construction equipment to

49

block a 2,600 km highway project that aimed to enhance the touristic potential of the region by connecting the high plateaus (*yaylas*)[1] of the East Black Sea Mountains to each other.

Halime Çakmak, mentioned in the previous chapter, echoes Mother Havva in an interview reported by Hilal Köse from the daily newspaper *Cumhuriyet*:

> This HEPP, it's the state's doing. "You cannot stand up to the state," they said. If you are the state, who am I? I am the people. . . . We used to be afraid of the gendarmerie. [But] we mothers give birth to them, raise them, and send them to military service.[2] Then one of them comes and tries to evict me from my father's registered land. . . . [At a protest to block the construction work] I took a stone and started to hit my head with it. At that moment, I saw that one soldier was crying. "Sister, the order comes from above," he said. Who stands above us, apart from God?[3]

This spirit is captured by a widely circulated photograph taken in 2011 during an anti-HEPP sit-in in Bağbaşı (EAR). The picture shows a group of local women, all wearing either headscarves or *niqabs*, crashing into the police shields to break through the barricade the police had made to guard the construction site.[4] Under normal circumstances, these women would not even share the same physical space with men, especially with men who are not their relatives or acquaintances. Here they clash with police officers, physically pushing male bodies with their own. Their bodily action, driven by an intense feeling of injustice, transgresses multiple borders of tradition, religion, and culture.

Women have assumed a leading role in the struggles for the environmental commons and environmental justice in the last few decades (see, e.g., Krauss 1993; Brown and Ferguson 1995; Naples 1998; Culley and Angelique 2003; Prindeville 2004; Stein 2004; Bell and Braun 2010; Unger 2012; Bell 2013). Many local movements against environmental degradation and dispossession across the globe, from the Green Belt Movement to Chipko Andolan, from Love Canal and St. James Parish to Standing Rock, have been driven by women. Women are also known to be more radical, courageous, and committed to grassroots environmental struggles than men, with their more cautionary attitudes (see, e.g., Garland 1988; Krauss 1998; Prindeville 2004; Bell 2013). The antihydropower movement in Turkey might not be categorized as a women-led movement per se, but women's visibility, commitment, and courage have been defining elements of the movement (Hamsici 2010; Kasapoğlu 2013; Kepenek 2014; *Radikal* 2015), especially in, but not limited to, the East Black Sea Region.

Hundreds of pictures and videos of women demonstrating in their traditional clothing of shalwars (*şalvar*), long skirts, and headscarves—carrying banners in protests, shouting slogans, keeping guard at the construction sites and blocking construction equipment, and physically confronting the military and the police—have been widely shared through news reports and social media, creating substantial public impact. Wherever women effectively mobilize themselves against HEPP projects and gain public visibility, the movement manages to cultivate strong public support and stop these projects in different regions of the country. It was a novelty to see these women—conventionally portrayed as hardworking but docile, confined to the private sphere of their homes—dominating the public image of a protest movement. It was the first time in Turkish history that peasant women entered the political scene of activism and protest on such an impressive scale. And it is safe to say that these women protesting to save their rivers have become ingrained in the collective memory of the people in Turkey and have had a significant impact on the outcomes of the movement.

This chapter aims to demonstrate how to operationalize the conceptual framework of critical feminist scholarship on gender and environment, taking women's anti-HEPP activism in the East Black Sea Region as an empirical case. The case can also be used to demonstrate the limitations of this scholarship for studying women's environmental activism. To overcome these limitations, in this chapter I start by developing a body-centered conceptual framework inspired by Merleau-Ponty and contemporary feminist theory and establish a novel perspective to study the relationship between gender and environment in the context of local environmental struggles. I develop this framework not as a purely theoretical exercise, but on the basis of the experiences, stories, and narratives of East Black Sea women who fight against the HEPPs in their villages and valleys. On this empirical ground, I will identify the potential contribution of the concept of the lived, phenomenal body to critical feminist scholarship

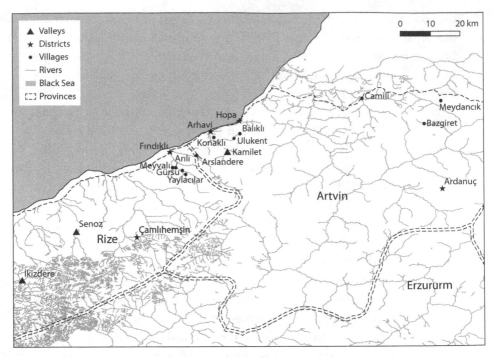

Map 2. The northeastern part of the East Black Sea Region, with villages and valleys mentioned in the book. Map by Serhat Karakayali.

on gender and environment, women's political agency, and environmental
activism. The chapter ends with a section that situates the lived body
within specific relations of power characterized by the patriarchal organization of everyday life.

GENDER DIFFERENCE WITHIN THE ANTI-HEPP
MOVEMENT: DISCOURSES, NARRATIVES, AND THE ACT
OF SAYING "NO"

When asked about how they identified their grievances and why they were
against HEPPs, the majority of the local men I interviewed in the East
Black Sea Region located the issue within a macropicture of complex
political processes, histories, and sentiments. In that sense, men often

talked about their take on what we can call *the politics of fresh water*. A typical conversation recorded as I sat at a table with seven or eight men in the village coffee house in Arılı (EBR) in the summer of 2013 illustrates the hegemonic narratives of the politics of fresh water. There emerged a strong consensus among them when Yusuf, in his forties, pointed out that HEPPs were just a cover for grabbing the fresh waters of the Turkish Black Sea Region. He explained: "Scientists say that global warming will be a huge problem in twenty-five to thirty years. Turkey will turn into a huge desert apart from the East Black Sea Region . . . ; that's why foreign powers want to get hold of our rivers under the guise of electricity production." Cemal, in his seventies, said, "No need to talk a lot—Armenia will inherit our lands." Refik, in his fifties, mentioned Israel and the United States as the big players in the game. Yusuf continued: "Remember the partition of Ottoman land after the First World War? They are trying to do what they could not do ninety years ago, by various means." When I asked, he confirmed that he was comparing the anti-HEPP movement to the War of Independence.[5]

The association with the War of Independence is also used to differentiate legitimacy from legality. Mehmet, the owner of the coffee house in which the group interview was conducted, said, "Didn't the Ottoman Sultan legally give our land to Western powers [referring to the Treaty of Sèvres]? But Mustafa Kemal and the Turkish people did not accept it. They ignored those legal documents, and they fought. If that struggle was wrong and unjust, let our struggle be wrong and unjust too." As this example illustrates, men's discourses and narratives merged notions of global warming and climate change with imperialistic, hegemonic struggles over fresh water and conspiracy theories, with national history and collective memory thrown into the mix. In other words, technical language and scientific information, provided mainly by the academics who support the cause, stands side by side with nationalist sentiments characterized by deep-seated fears and prejudices. For instance, men often talked about the marginality of the electricity that the HEPPs would produce. Hundreds of small-scale HEPPs, which would destroy the unique natural landscape of the East Black Sea Region, would only produce 0.4 percent of the country's energy needs, they argued, whereas the energy loss due to infrastructural problems such as inefficient transmission lines is around 27 percent, the implication being that the effort and resources would be put to better use improving the country's electrical grid. They mentioned alternatives to hydropower, suggesting that Turkey should invest in solar, thermal, and wind energy technologies if the aim is to produce renewable energy. Talking about the ecological damage that HEPPs would cause, men often used sophisticated language that involved terms such as *flora, fauna, endemic,* and *ecosystem* to explain the complex interconnections between water and plant and animal life.

Unlike men, most women do not refer to struggles over fresh water and foreign conspiracies, nor do they cite statistics or mention the War of Independence. Instead, women talk about their everyday sensory connection to river waters, their memories, and their sense of place: growing up with a view of the river, not being able to sleep without the sound of the river, the feeling of peace and relaxation when they put their feet or body into the river waters, the children who learn to swim in the river, the memories of their parents by the river.[6] The framing of the women's grievances and their opposition builds on a deep reservoir of geographical knowledge, sensory experiences, everyday interactions, and material practices, demonstrating that they do not need the mediation of the many references and analogies that men employ to express their grievances and legitimize their political agency. There are, of course, common themes and narratives shared by men and women, such as seeing the rivers as the central element of both the physical landscape and the geography of their everyday lives, perceiving their riverside villages as paradise on Earth, and emphasizing the rights of other living beings (animals, plants) and the uniqueness of East Black Sea ecosystems, landscapes, and riverscapes that are being violated by HEPP projects. The differences, however, are more striking than the similarities (Yaka 2019a).

Georgina Drew (2017) observes similar gender differences in the way men and women living in the mountains of Garhwal relate to the Ganges (Ganga) River and frame their religious motivations against the hydropower projects in their region. While men cite Hindu texts and scriptures to stress the religious significance of the Ganges, women refer to their less formal everyday practices and interactions with the river. Women's religious practices involve daily interaction with river waters as they begin their day by washing their faces, hands, and feet or enjoying a full-body dip.[7] Drew maintains that most men, as they are typically more educated than women, use standardized sacred chants during ceremonies of ritual bathing in the river. On the other hand, women merely speak "what is in their hearts while asking for guidance and blessings" (Drew 2017, 124), making their connection to the Ganga more intimate. Drew also stresses that women's enthusiastic, passionate activism is driven by their close connection to the Ganga. When one compares Drew's account of women's antihydropower struggle in Uttarkashi to countless journalistic observations of the gendered differences within the local anti-HEPP movements in the East Black Sea Region, similar patterns emerge. In both cases men refer to higher forms of authority or broadly recognized sources of knowledge (scriptural religious knowledge in India, scientific and political discourse in Turkey), while women draw on everyday practice and bodily experience.[8]

Journalist Özge Ozan articulates the same phenomenon with a slightly different accent in her news report on resistance against HEPPs in Ardanuç (EBR):

> Of course, we have also talked with local men. But when we spoke with women, we could not help but notice a clearly observable difference. Women talk like the water, like the river, gurgling, without calculating. There is no place for technical data, evidence, proof, or discourses developed to discredit the arguments of the adversaries in women's talk. But there is determination; there is resistance. Women say, ". . . we would rather die than give our rivers away," and they end it there. (Ozan 2010)

The last sentence of the quote reveals yet another aspect of the gendered difference: women's radical, bold, and committed activism. It has always been the women who brandish sticks, rocks, knives, and other "weapons" for journalists to see, posing proudly with them, implying that they will not hesitate to use their weapons against those who attempt to "steal" their river from them.

In the village of Yaylacılar (EBR), I recorded a conversation with a family in front of their house. One young woman was complaining about her mother, who had posed with a large, sharpened sickle for a national newspaper, implying that she would not hesitate to use it on whoever came to build a HEPP. In response to the young woman's concern that posing with a sharp object could be considered a criminal offense, her mother said, "Isn't taking our river a crime, Ayşegül?" Such scenes have been observed by quite a few journalists, researchers, and filmmakers, including Mahmut Hamisici (2010) and Yücel Sönmez, who interviewed women resisting not only HEPPs but also stone quarries, gold mines, and coal plants across the Black Sea Region. Sönmez quotes Ayşe Fettahoğlu, a fifty-six-year-old woman from the district of Tonya (EBR), who says, "They should consider that women here know how to fight, how to use a gun. They should not force us to do these things. . . . We are on the front line as women. Without us, there would not be a struggle anyway. . . . I am not afraid of anything" (*Radikal* 2015). He also quotes sixty-seven-year-old Rukiye Bicil from the village of Engiz (EBR), who says, "Men do not have the courage we have. If it were up to them, the resistance would slacken." Forty-three-year-old Sabire Tatli from Solaklı Valley (EBR), on the other hand, states:

> We were detained and taken to the police station. They dragged my 80-year-old mother around. They want us to be scared. Most men gave up. They [the police] caused discord between them; they scared them. But they do not know us, the Black Sea women. They do not know our courage. . . . We know what we are fighting for. They cannot trick us. (*Radikal* 2015)

These statements echo many others around the world, such as accounts of Appalachian women fighting against mountaintop-removal mining who claim that "men are too chicken" (Bell 2013, 91; see also Garland 1988; Kraus 1998; Prindeville 2004). Courage is not necessarily a personal trait or a group identity (such as in the above example touting the courage of Black Sea women); it is instead a product of a strong sense of purpose, determination, and commitment. Courage, thus, very much depends on the sense of injustice inflicted and the relative urgency of the cause. Dai Qing, a Chinese journalist who is known for her activism against the Three Gorges Dam, states in her Goldman Environmental Prize speech: "The highest expression of dignity can be summed up in the single world 'No!'" (quoted in Nixon 2011, 150). Indeed, all social movements boil down to the act of saying "no" in the last instance.

The act of saying "no" is loaded with gendered meanings. Let me use a fascinating short film titled *Le Reprise du Travail aux Usines Wonder* ("Resumption of Work at the Wonder Factory")[9] as a cultural frame of reference to explore the gendered meanings of saying "no." On June 10, 1968, a group of workers and activists gathered in front of the Wonder factory, a battery-producing plant in Saint-Quen, a northern suburb of Paris. After three weeks of strike and occupation, employees voted to end the strike. The film opens with a woman, a worker in the factory, speaking to two well-dressed male representatives of the local trade union in an overwhelmingly male crowd. While the trade unionists try to frame the whole thing as a victory and remind her that the decision has already been made and she needs to return to work, the woman continues to resist. "No. I am not going back. No. I am not going back. No!" she says and adds, "We got nothing. One week of holidays. I am not going back to that dump."

The woman shows her emotions—anger and sadness—crying while talking about what a horrible place the factory is for the workers and how the supervisor has rigged the vote. "I am not going back. Not with this boss," she says. She could not wrap her mind around how on earth the minuscule concessions the supervisor had made to end the strike could be framed as a victory or even a significant step toward it. A few minutes into the conversation, a young male activist says something to support the woman, and the trade unionists turn to him and continue to discuss with him. The camera angle also changes to follow the conversation while the woman moves out of the center of the frame. Whenever she says something, from the corner now, someone says, "Silence, silence." The film ends with workers, primarily women, passing through a door into the factory one by one, but we also see a woman, not the one we saw at the beginning, who leaves instead of going in, saying, "I don't want to go in. I don't want to."

The film depicts the (typical) attempt to silence "loud," "emotional," and "irrational" women in the particular context of strike and occupation in Paris in 1968. This act of silencing women is not unique to 1968, France, or labor struggles. It is all too familiar to scholars of women's environmental activism (see, e.g., Seager 1996; Naples 1998; Unger 2012; Bell 2013). There are also a few other aspects of the film that stand out. The first is the woman's loneliness within a group of men who decide to reconcile with the supervisor and end the resistance, clearly without her approval. The only support she got was from the young male activist who does not work in the factory and from one other woman who refuses to go back to work at the end of the film.

The loneliness of the woman in the film reminded me of another young woman, Ayşegül from the village of Konaklı (EBR), who told me about the emergence of strong local resistance against the Kavak HEPP project (the planned location of which was to be between Konaklı and Kemerköprü villages in the district of Arhavi) and its eventual failure. As I gathered from my interviews with Ayşegül and other residents of Konaklı and Kemerköprü, the resistance against the Kavak HEPP (known locally as the "Crazy HEPP"[10]) was broken by a series of interventions by the company undertaking the project (MNG Holding), whose owner is also from Arhavi. The company representatives negotiated with the local "leaders" of the movement, a group of local men, and agreed to make minor changes in the project to mitigate its socio-ecological impacts. It is rumored that some of these men also derived personal benefits ranging from cash to job offers and overpriced leasing deals.[11]

In the end, those men changed their discourse drastically. Sabri, a local bank manager I interviewed in the summer of 2013 in his office in the Arhavi town center, acknowledged that the HEPP project would cause social and ecological harm, but rationalized it with a series of sometimes conflicting arguments ranging from the need for energy and investment to the goodwill of the HEPP company to the assertion that nature had "already" been despoiled by the ongoing construction. "It would be constructed anyway," he said. "There is no point in resisting." He also said that "everyone" was satisfied with the project after the amendments. Ayşegül, on the other hand, was not convinced. "I will say no, even if everyone else says yes." Indeed, in the end it was the women, who had organized themselves under the name "Women Sparrowhawks" (Kadın Atmacalar), who kept resisting the Kavak HEPP and other HEPP projects in Arhavi. Nevertheless, the construction of the Kavak HEPP continued, despite the fact that a local court twice refused to approve its environmental impact assessment report. After construction of the HEPP, the Cihani River dried up. "We used to jump in the river from trees; now, the water barely reaches our ankles. Our hearts bleed," said Nazlı Demet Uyanık, a young member of Women Sparrowhawks (Sputniknews 2018).

Returning to the film about the Paris factory strike, an underlying theme that the scene seems to deal with is the gendered contradiction between the determination and stubbornness of the woman, who appears to be existentially opposed to reconciliation, and the accommodationist arguments of the men, who portray the relationship between workers and the capitalist supervisor as one of conflict, negotiation, and reconciliation. The scene provides a critical representation of a dominant narrative: that men behave responsibly and reasonably while women act stubbornly and emotionally. However, many social protest movements owe their public impact and (relative) success to their determination, stubbornness, and mobilization of intense feelings and emotions. Rukiye Bicil's remark about how the resistance would slacken if it were left up to the men (quoted above) illustrates the significance of women's courage and determination to sustain the anti-HEPP resistance in the long term. Let me encapsulate her argument through a conversation I recorded in the village of Gürsu (EBR) in the summer of 2013.

The district of Fındıklı, in which Gürsu is located, is a symbolic place for the anti-HEPP struggle, as the movement has remained strong, spirited, and vigilant there for more than a decade and has managed to block each and every HEPP project so far. The conversation below demonstrates the role women's radicalism and commitment have played in keeping the movement so strong and vigilant. As I was talking to a group of middle-aged and older women on a doorstep in Gürsu, two men came and joined the conversation. After questioning my "secret intentions" (a roundabout way of asking if I was an agent of a company), one of the men, Hüseyin, who was an engineer working in a large city, started to explain his take on the issue (Saliha, Melahat, and Ayşe are three women who got into a discussion with him; ÖY stands for me):

HÜSEYIN: The village can earn money from a small-scale HEPP. We could build one ourselves, produce our energy, and sell the rest [to the state]. It would also bring jobs to our villages. But the laws should be regulated or amended so that local people would benefit from the process, not just private companies.

SALIHA: We do not let the law [intervene]. No one should come to us with a HEPP project. We do not want it.

HÜSEYIN: Wouldn't it be good to produce electricity and get money from the state? In addition to what you earn from the tea harvest?

MELAHAT: Are you on the side of selling the river? Are you?

HÜSEYIN: No, I am saying that you would earn money. I am saying only that.

SALIHA: We don't want the money.[12] We can pay for the electricity we use. We do not want the state to interfere with us, with our river.

MELAHAT: No one should come to sell it![13]

SALIHA: Whoever wants to sell our river should sell their own families instead. We are content with what we have. Some people want it, we know that! [Said in a threatening tone toward Hüseyin]

MELAHAT: Are you on their side?

SALIHA: Our children grew up here. We go to the river every day.

HÜSEYIN: I understand what you are saying. I am on your side.

SALIHA: They already ruined the sea. It got polluted. We cannot go to the sea anyway.[14] We want to swim in our clean river. What do they want from us? They should not come here. I don't want to see them [the company people] here.

ÖY: I don't think you would give it to anyone. I am talking to women from Yaylacılar [located higher up in the valley] to here; they all said the same thing.

MELAHAT: No, no, no! We would not!

SALIHA: My children and grandchildren are in the river every day.

ÖY: I saw some kids playing in it, very beautiful.

SALIHA: It is also very clean, you know. And there are waterfalls in this small river, can you believe it?

MELAHAT: But Hüseyin has spoken as if he is on the side of selling the river.

AYŞE: For him, we should sell the river, and then what? Leave this place altogether?

MELAHAT: We should construct a plant and get the money, huh? We do not want anything. We are content with what we have. We are happy with our village, our river.

Seeing that his ideas were not popular, Hüseyin grew silent toward the end of the conversation and left. I suspect that the strong reaction of the women discouraged him from voicing his compromising ideas in public again. This conversation resonates with journalist Özge Ozan's observations in Ardanuç, another town in the same region:

> During our conversations with women, we do not hear sentences such as, "The country needs the energy" or "How can one oppose the state?" Nobody says, "But HEPP projects would bring new jobs." . . . Job promises and bribes

are not given to women anyway, and they do not care about the electricity argument as they have lived without electricity in the past, but they cannot imagine living without the river. (Ozan 2010)

INTERPRETING GENDER DIFFERENCES IN RELATION TO ENVIRONMENTS IN THE EAST BLACK SEA REGION

My ethnography and many journalistic accounts of the anti-HEPP movement in the East Black Sea Region (see esp. Ozan 2010; see also Hamsici 2010; Kasapoğlu 2013; Kepenek 2014; Yavuz and Şendeniz 2013 *Radikal* 2015) indicate "real, not imagined, gender differences in experiences of, responsibilities for, and interests in 'nature' and environments" (Rocheleau, Thomas-Slayter, and Wangari 1996, 2). And if we are not content with essentialist explanations for gender differences—that women are essentially more resistant and men are more compromising, or women have an essential affinity with nature—if "these differences are not rooted in biology per se" (Rocheleau, Thomas-Slayter, and Wangari 1996, 2), we need to problematize them. Let us consider the gendered nature of receiving benefits, bribes, and job promises, which is deeply related to the gendered division of labor and allocation of responsibilities. "Men can be tempted to take money from the companies, whereas for women, protecting the rivers for the next generations is more important," two elderly women in the village of Aslandere (EBR) told me when I asked them why women were more committed to preventing HEPP projects.

Bina Agarwal explores the same issue in her influential article, "The gender and environment debate: Lessons from India":

Women's active involvement in the Chipko movement . . . [took place] . . . even in opposition to village men due to differences in priorities about resource use. Time and again, women have clear-sightedly opted for saving forests and the environment over the short-term gains of developmental projects with high environmental costs. In one instance, a potato-seed farm was to be established by cutting down a tract of oak forest in Dongri Paintoli village. The men supported the scheme because it would bring in cash income. The women protested because it would take away their only local source of fuel and fodder and add five kilometers to their fuel-collecting journeys, but cash in the men's hands would not necessarily benefit them or

their children. . . . Once again this points to the association between gendered responsibility for providing a family's subsistence needs and gendered
responses to threats against the resources that fulfill those needs. (Agarwal
1992, 147–8)

Agarwal maintains that gendered priorities and responsibilities are central
to gendered attitudes toward the environment. Similarly, Unger (2012) and
Bell (2013) discuss how women prioritize health and safety (of the family
and children) while men remain focused on economics, supporting toxic
industries as long as they provide jobs, as in the case of African American
women's struggle against the Shintech PVC Plant in St. James Parish,
Lousiana (Unger 2012, 196). US-based environmental justice scholarship
has demonstrated how women's environmental activism is closely affiliated
with their (gendered) role as mothers and pillars of the family through various case studies of struggles against toxic waste (see, e.g., Pardo 1990;
Krauss 1993, 1998; Brown and Ferguson 1995; Stein 2004; Verchick
2004). Besides women's gendered roles as mothers and pillars of the family,
their position as guardians of community, land, and culture gains particular
weight when it comes to the struggles of Indigenous populations around
the environmental commons, especially in the US-based literature (see, e.g.,
LaDuke 1999; Prindeville 2004; Jenkins 2015).

Women's environmental activism in the Global South, on the other
hand, is discussed primarily in relation to livelihoods "that are directly
linked to the use of natural resources, such as water, fish, forests, and wild
animals" (Cruz-Torres and McElwee 2017). Besides and beyond gender
roles, the gendered division of labor emerges as a central variable here, as
women are typically responsible for sustaining livelihoods. Livelihood is a
concept that blurs the conventional boundaries between production and
social reproduction. It involves a series of everyday activities (in addition
to housework) that fall within women's sphere of responsibility, such as
subsistence agriculture and animal husbandry, fetching fuel and fodder,
and securing water for irrigation, drinking, and household use—activities
that are seen as an extension of household work. Because of their everyday
material practices to sustain the lives of their families, the argument goes,
women are connected to the nature and environments on which their livelihoods depend in a way that men are not. This emphasis on gendered

social relations, material practices, responsibilities, and division of labor has been central to the evolution of earlier ecofeminist[15] and WED (women, environment, and development) approaches to ecological and environmental feminism, critical ecofeminism, and feminist political ecology.[16]

Now we can turn to the East Black Sea case in light of the feminist literature on gender and the environment. Drawing on this literature, it would not be difficult to detect essentialism in the stereotypical portrayal of Black Sea women as strong, courageous, and devoted, as employed in many journalistic and some academic accounts of HEPP movements to explain women's commitment and radicalism (see, e.g., Yavuz and Şendeniz 2013; Kasapoğlu 2013). Although many local movements against HEPPs use this stereotype strategically and instrumentally (Işıl and Arslan 2014), it, like all stereotypes, conceals certain aspects of the complex social fabric of everyday life while overemphasizing others. Even though the nature of their hard agricultural labor makes them physically strong and reserves for them a legitimate social position within the community, women are still economically and socially vulnerable in the face of the patriarchal organization of social life. Hence, instead of relying on stereotypical, essentialist portrayals of East Black Sea women, one should focus on positionality, gender structures, and the socio-spatial organization of everyday life that positions men and women differently and, consequently, enables them to generate different experiences and bodies of knowledge.

The organization of gender relations in the typical rural household of Turkey's East Black Sea Region conforms to Deniz Kandiyoti's definition of classic patriarchy, characterized by extended family structures, examples of which can be found in North Africa, the Middle East, and South and East Asia (Kandiyoti 1991). As in many other parts of the world (see, e.g., Eaton and Lorentzen 2003; Federici 2012), women in the East Black Sea Region are responsible for domestic and agricultural work, but they have no claim to their family's inheritance. As in other regions of Turkey, women's participation in the workforce is low, except for the agricultural sector.[17] Agricultural work is seen as an extension of housework; as such, it is strictly coded as women's work.

Tea is one of the principal monocultural agricultural products of the region, along with hazelnuts, which are grown in the western part of the

East Black Sea Region, especially in the Ordu and Giresun Provinces. In the eastern part of the region, though, in the Rize and Artvin Provinces, where I conducted my fieldwork, tea dominates agricultural production. Fruit and vegetable agriculture also exists, but only to meet household needs. Unlike in India and China, tea is typically not grown on large-scale plantations, but in smaller, family-owned fields. Women work as unpaid family workers, planting, fertilizing, pruning, and harvesting in these fields, which are legally owned by men.[18]

East Black Sea women are typically depicted with large, heavy baskets on their backs, in which they carry harvested tea leaves and other agricultural products, fertilizer and tools to be used in the field, bushes to be burned, and sometimes even children (see, e.g., Şendeniz and Yıldırım 2018). These baskets are associated with women in the region; it is regarded as odd, and sometimes even shameful, for a man to carry a basket (Bellér-Hann and Hann 2000; Karaçimen and Değirmenci 2019). It is not unknown for men to help women with the agricultural work, but it is a relatively rare occurrence.[19]

Local men mainly work as paid workers outside the village, either in the district and town centers (as fishermen, drivers, shopkeepers, and workers or civil servants for CAYKUR, the General Directorate of Tea Enterprises, which is the state-owned tea-production company). Some leave the area for work in Turkey's larger cities, or even abroad. During harvest season, they tend to return to their villages, primarily to undertake the transportation and transaction of the harvest. Hence, it is men who receive the money in the end. Men are also more mobile; they visit the town centers and other cities more frequently than women and thus encounter the wider public and the state more often than women. It is primarily men who are in contact with institutions such as their local bank, the tea factory, the local branch of the Ministry of Agriculture, the municipal administration, and the district governorate. Men must also serve as soldiers within the notoriously repressive institutional structure of the Turkish Army. Consequently, they are confronted both with the bureaucratic rationality of the state and its violent, repressive apparatuses more often than women.

These close encounters with the broader public, and especially with the state apparatus, are constitutive of men's subjectivization processes in

Turkey. These encounters are also reflected in men's framings of the anti-HEPP movement, which are conditioned by their quest for legitimacy in the eyes of a wider public and the state. This quest for legitimacy leads them to place the antihydropower struggle within a broader political frame of national resistance to imperialism and to use technical language in support of their position. Men's compulsory engagement with the army is particularly important here to understand the socialization and subjectivization of men, which also shape their discourses and political agencies in the context of the anti-HEPP movement. Military service of either six months or one year is compulsory for all men who are Turkish citizens.[20] Compulsory military service functions as a disciplinary practice of citizenship through which young men from all over the country face the state's repressive apparatus both as subjects and objects of state violence. It also engenders certain narratives of nationalism and masculinity, the effects of which go far beyond the military as an institution (see Altınay 2004). Such narratives are dominant in what Bozok defines as "the hegemonic masculinities of the East Black Sea Region," which mixes conservative and nationalist discourses with cliché qualities like toughness and zeal (Bozok 2013).

This notion of hegemonic masculinity is helpful in exploring men's framings of the anti-HEPP movement, especially their urge to invoke nationalist narratives. The same structures, institutions, and practices that inform hegemonic masculinity, however, also indicate its limits, as the male subject's masculinity is permitted and encouraged only when it is subjected to and in cooperation with the state apparatus. The awareness of those limits makes men more cautious than women in expressing their opposition to HEPPs, preventing them from employing a more radical discourse as women do. Instead, they generally refrain from proffering such threatening messages, especially against the state. The BBC Turkish concisely depicts this striking contrast between the discourses and framings of men and women in the fifth episode of their video reportage series on HEPPs in the Black Sea Region (*Karadeniz'de HES'ler*).[21] The video, which depicts the struggle in Çiftlik, a village in Korgan District (EBR), starts with the voice of a middle-aged man who, as we learn later in the video, is called Kemalettin Topuz, saying, "What we want is to protect our nature. We do not oppose anyone, not the state nor our soldiers."

A woman's voice immediately follows. It belongs to a middle-aged woman, Gülseren Topuz, who, judging from the shared surname, is most likely Kemalettin Topuz's wife or a close relative. She says, "They cannot [do it]. I will die for it. I won't allow it."

The difference in tone and sentiment is indeed striking, and BBC Turkish almost certainly edited the video to underline this difference. The cautious, reassuring language of Kemalettin Topuz reminded me of men like Metin and Cengiz, with whom I spoke in the village of Arılı (EBR), where women did not hesitate to talk about dying and killing for the cause. When I asked them, separately, what would happen if the HEPP company came with the gendarmerie and attempted to construct the HEPP despite the court decision to cease construction activities in the Arılı Valley, they could not give me a clear answer. Eager to fend the question off, they both ruled out the possibility of a physical confrontation with the state's armed forces. "The state would not do anything to harm people," said Cengiz, even though he knew that precisely that had happened in many other villages and valleys. Faruk, a middle-aged teacher from the same village, used another strategy that manifests men's caution. He was sitting at a table outside the village coffee house when I approached him, introduced myself, and asked him what he thought about the HEPPs. He told me he supports them. Surprised by his answer, I wanted to sit and pursue the conversation. He then asked me about my position. When he realized I was sympathetic to the anti-HEPP movement, he told me he was also against it. "I was testing you," he said, citing his hesitation to reveal his position openly, as (unknown) people might report him to the authorities.[22]

BEYOND THE EXISTING LITERATURE: EMBODIED EXPERIENCE BETWEEN DWELLING AND LIFEWORLD

As discussed in the empirical context of the East Black Sea Region above, the existing feminist literature on gender and environment is indeed helpful in going beyond essentialist stereotypes, revealing the structural and institutional factors, everyday practices and organization of daily life, and gendered division of labor and responsibilities, all of which shape gender differences in a given spatiotemporal context. The East Black Sea case, however, belies a very central argument of that literature: it is because women are responsible for sustaining livelihoods, the argument goes, that they are more threatened by and reactive to enclosures of environmental commons, which are framed in the literature as "natural resources" on which the livelihoods that women are trying to sustain directly depend.[23] This has been the case especially for water, which is often framed as a vital "natural resource"; accordingly, water struggles are presented as livelihood struggles par excellence, motivating women to become highly mobilized (see, e.g., Bennett, Dávila-Poblete, and Rico 2005; Braun 2008; Buechler and Hanson 2015; Levin 2016; Rusansky 2020). And wherever water is seen as more than a resource, its cultural value is recognized in the context of religion, belief, and spirituality.

In most of the East Black Sea Region, however, "livelihoods" do not depend on the river, as discussed in the previous chapter.[24] Even though religious notions inevitably influence the conceptions and experiences of river waters in the region—e.g., "God's water"—water is not understood as particularly sacred or associated with belief (with the exception of a few isolated examples, see footnote 29 in the previous chapter). The empirical case of the East Black Sea Region, then, leaves us with a crucial question: Why are women so committed to protecting the rivers even though river waters are not used for immediate economic purposes or perceived as "sacred"? Responding to this crucial question requires going beyond the existing feminist literature on gender and environment and bridging different disciplinary conversations. Let us begin by going back to the phenomenological tradition, primarily phenomenological humanist geography, which provides conceptual tools that correspond to the concept of livelihood, but also gives us clues for moving beyond that concept.

Two central concepts of phenomenological geography are *dwelling* and *lifeworld* (see, e.g., Buttimer 1976; Seamon 1979; Buttimer 1980). Dwelling, a concept developed by Martin Heidegger, refers to the processes of placemaking, that is, human practices of building, cultivating, and creating a place to live (Heidegger 1971; see also Casey 1993, 2001; Ingold 2000; Wylie 2003; Harrison 2007; Donohoe 2014; Ash and Simpson 2016). Heidegger's appeal to scholars of place, landscape, and environment, despite his very dark record of Nazi affiliations, is due to the conceptual capacity of dwelling, which indicates a subject's essential situatedness—building on the claim that all existence is emplaced (see, e.g., Larsen and Johnson 2012b). Maintaining what Wylie (2003) articulates as "agent-in-its-environment," Heidegger's work "intimately interlocked" place and the self "in the world of concrete work" (Casey 2001). As is stressed repeatedly in the contemporary geography literature, neither the subject nor the environment precedes this relationship; they both emerge within this very relationality through the everyday practices of dwelling (see, e.g., Harrison 2007; Rose 2012).[25]

The "work-world" is central to dwelling, and it is the habitual patterns and micropractices of routine labor that interlock the self intimately to places and environments. This is precisely the argument the feminist environmentalism of Bina Agarwal and the feminist political ecology that followed her lead have been making over and over in the last few decades. Women who live in "natural environments," mostly in rural areas, are tied to "nature"—to lands and mountains, valleys and forests, lakes and rivers—through their habitual patterns of (mostly unpaid) work, which is the central aspect of dwelling. In that sense, dwelling is surely central to people's—particularly women's—relationship with nature, as discussed in the previous section. The hardships of dwelling in the East Black Sea Region are incontestable. Building a homestead requires a great deal of work, as it is difficult to find flat surfaces, and people have to use ropeways to transport goods across the rivers within deep valleys. Agriculture is also challenging, as most of the land is on steep slopes. The need to overcome these hardships makes dwelling practices central to everyday life.

Another important concept from the conceptual vocabulary of phenomenological geography is *lifeworld*. The concept of lifeworld, which originated in Husserl's work but is employed by many others, including Jürgen Habermas, designates a world of lived experience (see, e.g., Schutz and Luckmann 1973; Buttimer 1976; Seamon 1979; Habermas 1987; Ihde 1990; Dahlberg, Drew, and Nyström 2001). It has been a central argument of phenomenological geography that our routine interactions with this shared socio-material world, which are the basis for all experience, are taken for granted and not usually considered of analytical value (Buttimer 1976; Ash and Simpson 2016). The formation of self and subjectivity, however, is very much embedded in these routine interactions and experiences of everyday life.

If we relate these two concepts, dwelling and lifeworld, in the empirical context of the East Black Sea Region, we can say that the structures of dwelling, that is, the gendered division of labor that typically puts women in a disadvantaged position in many ways, also puts them in a close(r) relationship with their immediate environments. It requires women to work in the fields, shaping their world of lived experiences, which involves everyday interaction with river waters. Hence, the gendered structures and practices provide the material basis for the unfolding of everyday experiences. But these experiences are not limited to or dominated by the instrumentality of the work-world; they involve intimate bodily relations and connections between human and nonhuman beings. Furthermore, the value or "worth" of an environmental entity and one's experiences of that entity are not bound up with its instrumental use within the structures of dwelling (in the case at hand, the use of water in agriculture, for instance). The East Black Sea case indicates the close affinity between the value and worth of a nonhuman, environmental entity and its position within our world of lived experience, that is, within our lifeworld.[26]

Of course, it is not only women for whom rivers and river waters are central to the lifeworld in the East Black Sea Region. Men also grow up and live in the same villages and valleys, accumulating experiences and memories of river waters all their lives. But still, as the everyday routines and life courses of men and women differ considerably, as discussed briefly in the previous section, the habitual ways in which men interact with river waters also diverge. It is not the case that men are not connected to their environments. Still, the degree and quality of that connection differs between men and women, as connectivity with one's surroundings and one's lifeworld, in general, is shaped by gendered relations, institutions, and practices.[27]

CORPOREAL FEMINISM, MERLEAU-PONTY, AND THE
LIVED BODY AS A MISSING LINK TO RELATE GENDER
AND ENVIRONMENT

There is a tendency, even within progressive, feminist, and environmentalist circles, to see natural environments as an aggregate of resources.

The concept of lifeworld introduced above, when combined with the concept of dwelling, points to the possibility of going beyond seeing natural environments as an aggregate of resources, and understanding them instead as a dynamic totality of organisms, processes, and networks we relate to, live with, and, ultimately, are composed of: as our lifeworld. Combining dwelling and lifeworld also requires considering experience in a broader sense, produced through material dwelling practices and everyday interactions with our lifeworld. Experience is, as decades of post-structuralist scholarship have taught us well, a much-contested concept (see, e.g., Scott 1991). It is, however, indispensable, not as a direct route to an uncontested truth, but as our primary mode of producing knowledge through our intimate connections with our lifeworld, as shaped by power and discourse (see, e.g., Kruks 2001). It is due to its irrefutable relationship with knowledge that experience has been a central concept for feminist epistemologies (see, e.g., Harding 1987; Taylor 1998; Alcoff and Potter 2003; Harding and Norberg 2005). What is often overlooked, however, is that when we talk about experience, we necessarily imply bodies and embodiment because we live in this world as bodies, and we experience the world through our corporeal connection to it.

We experience and relate to our natural environments through our bodies. Even though it is fairly obvious that material practices and materiality, experience, and the gendered division of labor necessarily involve women's bodies, a body-centered approach to gender and environment is largely absent from the canonical literature, with perhaps the exception of environmental health studies.[28] When this involvement is mentioned, it remains mostly superficial, reducing bodies to simple vehicles of work and livelihood and articulating them as "theoretical, discursive, fleshless" (Longhurst 2001). I believe the dominance of Foucauldian and post-structuralist accounts within the feminist literature on body, agency, and subjectivity has something to do with this superficial treatment of the body as a surface on which power and discourse act. A more profound understanding of the body, in terms of its perceptive capacities and its (trans-)corporeal connection with the environment, is still waiting to be developed in the field of gender and environment. Such a task requires what Elizabeth Grosz (1994) aptly calls "corporeal feminism," a profound engagement with certain variants of feminist theory, which is what I aim to do in this section.

In the aftermath of the so-called cultural turn, the 1990s witnessed a growing conceptual interest in materialities, things, bodies, objects, organisms, and networks. Taking the post-structuralist critique seriously but aiming to go beyond it, these new materialist theories of the social, from actor-network theory to nonrepresentational theory, attempted a relational articulation of materiality and discourse, as well as of nature and culture. Within this broader "material turn," it was mainly the feminist theorists, coming primarily from Spinozan/Deleuzian and phenomenological traditions, that radically reframed the body, as well as nature and matter, as active and dynamic, as generative, formative, and agential, and as indefinitely and unpredictably open to transformation and change (see, e.g., Grosz 1987, 1994; Gatens 1988, 1996; Braidotti 1991, 1994; Kirby 1997; Bray and Colebrook 1998; Weiss 1999). This line of thought has been instrumental in the formation of contemporary approaches, such as the new materialist feminisms in the 2000s.

What made corporeal feminism influential was its positioning of the body at the center of the feminist project of disintegrating Cartesian (gendered) dualisms. The understanding of the body as *res extensa,* literally an extension of the thinking substance (*res cogitans*), the separation of the subject from its bodily dimension, and the primacy of the former over the latter is one of the main pillars of modern philosophy from Descartes to Locke and Kant: "Modern thought places body under the rubric of the object. The body is what the subject recognizes inside itself, as different from itself" (Esposito 2015, 108). By attacking, conceptually, the idea of the passive and inert matter-body-nature and by refusing to line up with either nature or culture or to reduce one to the other, corporeal feminism truly destabilized the nature/culture binary. Consequently, corporeal feminism opened up a new conceptual pathway that avoids the pitfalls of both reproducing the Cartesian dualities in the form of materiality vs. representation (Bray and Colebrook 1998)—as in the sex vs. gender binary[29]—and merely reversing the mind-body hierarchy, as certain essentialist feminisms do. Instead, the body is framed as the location of the self, as the condition of knowledge and consciousness, and as the very "stuff" of subjectivity (see, e.g., Rich 1984; Grosz 1994; Young 2005).

It is in this sense not surprising that Merleau-Ponty has been one of the main inspirations for feminist scholars who have been working on body and subjectivity. He is, after all, the one who firmly anchored perception, consciousness, and experience in the phenomenal, lived body and, thus, changed the conceptual track of phenomenological thought. He maintained that our bodies are our horizons, our locus of perception and experience. And—through perception and experience—knowledge, consciousness, and political agency are anchored in the lived bodies. We not only experience and perceive; we also come to know the world and ourselves and act within and toward our human and nonhuman environments, always and necessarily as bodies.

Feminist theorists have famously criticized Merleau-Ponty's anonymous framing of the body and his neglect of sexual difference (Irigaray 1993; Butler 1989; Grosz 1994; Young 2005);[30] but Merleau-Ponty remains a central reference, especially for feminists who are looking for conceptual alternatives to viewing the body "as a passive product of cultural crafting" (Oksala 2006, 225). Merleau-Ponty's concept of the "body-subject" and his phenomenological conception of the "lived body" came in handy for the task. The conception of the phenomenal, lived body was employed to differentiate the body-subject from the physical, biological body, which has long been framed not only as inert and passive, but also as the very source of women's subordination, as a prison for women made of flesh and blood. From a phenomenological perspective, the lived body is not a prison, but our mode of *being-in-the-world*.[31] I claim that the lived body is a vital conceptual tool for fleshing out the main pillars of critical feminist scholarship on gender and environment. Let me present this argument by identifying two ways in which Merleau-Ponty's conception of the lived body as body-subject aids us in exploring this relationship.

1. THE LIVED BODY IS THE BODY IN A SITUATION.

It is "our horizontal and vertical anchorage in a place and a here-and-now" (Merleau-Ponty 1964, 5) that "opens me up to the world and puts me into a situation there" (Merleau-Ponty 2012, 168). It is thus not a generic body but a situated one, located within a specific spatiotemporal and sociocultural context. This idea of the subject as situated, both constrained and enabled in its very situatedness, has been central to feminist epistemologies, from standpoint theory to posthumanist variants (see, e.g., Haraway 1988; Braidotti 1991; Harding 2004). Framing the subject as embodied and situated is crucial to understanding how women relate to their environments. This relationship is first and foremost a corporeal one, as women live, work, and recreate within these environments, perceiving, moving, and acting as corporeal beings. And it is not an abstract, general, or generalizable relationship; rather, it is constructed between specific—differently sexed, raced, aged, abled—bodies of women and particular environments and bodies of nature, be they land in a Native American reservation in the United States, mountaintops in the Andes, forests in India, or rivers in Turkey.

The situated subject, as spatially and historically located and embedded within specific social, cultural, and sexual relations of power, has been the feminist response to the prevalence of an abstract and disembedded subject. Haraway's notion of the "god trick," which targets this notion of a disembedded subject who can see everything from nowhere, is well known. Less known is a similar concept of Merleau-Ponty's: *pensée de survol* ("high-altitude thinking"; see, e.g., Merleau-Ponty 1968, 73), which addresses precisely the same notion. Lived body, or the material-semiotic body, as Haraway would put it, is the antidote to the god trick and high-altitude thinking, as it is the body that situates the subject within a world of encounters. We cannot jump out of our skins to live as pure, abstract, unconstrained subjects. Self is located in the body (Rich 1984) and is situated within the world as a body, as we cannot but live in the world "as flesh," as material bodies with certain physical, sensory, and motor capacities.

To understand the dynamics of women's environmental activism, to protect the rivers in the case at hand, it is essential to attend to physical, motor, sensory, and affective capacities of the lived body and how those capacities are mediated and manifested in different spatiotemporal and sociocultural contexts. In the case of the East Black Sea Region, women's corporeal connection to river waters is established through their everyday sensory experiences—touching, seeing, hearing, and tasting the river waters. This everyday sensory relationship creates affective responses such as pleasure, joy, and relaxation. It is thus the perceptual capacities of the body that connect us intimately to our environments. More often than not, though, these perceptual capacities are effectuated within certain habitual dispositions that are structured and regulated by the socio-spatial organization of daily life. Geographical and cultural configurations of patriarchal power dynamics and the gendered division of labor play a central role in the formation of these habitual dispositions. For instance, in a region where the lines of demarcation between inside the household and outside the household are stricter, where the natural landscape is not coded as an extension of the household, women would probably not be spending large parts of their everyday lives out in nature. Consequently, they would probably be unable to establish an intimate corporeal connection with river waters. Lacking such a connection, women might not become very committed opponents of HEPP projects.

Not only are the contexts in which the body perceives, moves, and acts shaped by spatial formations, social relations, and cultural habitats, but the bodily capacities, orientations, and comportments themselves are as well. Let us take the female bodily comportment Iris Marion Young describes in her famous essay "Throwing Like a Girl." With reference to Merleau-Ponty's work, she claims that female bodily intentionality is defined not with the bodily "I can," as Merleau-Ponty puts it. It is rather inhibited as "the feminine body underuses its real capacity, both as the potentiality of its physical size and strength and as the real skills and coordination that are available to it" (Young 1980, 146). She goes on to argue that "feminine bodily existence" is in "discontinuous unity with both itself and its surroundings" and feels more at home in enclosed, inner spaces, different from males, who are more comfortable in outer spaces (Young 1980, 147). I wish she could have seen the East Black Sea women. Contrary to what she describes, women of the region are bodily strong and confident; they perform bodily actions—carrying, pulling, pushing, and throwing—skillfully, with the ease of habituation. Moreover, as opposed to what Young describes, they feel most at home outside, in the unique natural landscape of the region.

The case of East Black Sea women illustrates that a historically and spatially specific experience of embodiment cannot be associated with "feminine bodily existence" as "typical of all women at all times" (Chisholm 2008, 11). We could relate this point to the feminist discussion of the anonymous body. From such an empirically informed perspective one could maintain that the anonymous physical body is not a uniform, static, universal foundation of subjectivity; it is instead "a dimension of intersubjectivity that we cannot isolate or localize" (Oksala 2006, 222). In other words, the anonymous body does not *determine* the personal or the subjective, but it does "underpin" them (see Merleau-Ponty 2012, 168). It indicates certain potentialities and capacities—of movement, perception, and sense constitution—the limits of which are unknown to us. Those potentialities, however, emerge through the lived body's perceptual and habitual engagement with the world (Coole 2007), always actualized very differently within historical, spatial, and intersubjective contexts (see Oksala 2006). In this sense, the lived body is always a process of *becoming*—it is never uniform, static, or identical (see Stoller 2000). To explore our material and corporeal relations to our environments, it is thus essential to attend to (perceptive, sensory, affective) capacities, orientations, and habituations of lived bodies, as they are shaped within specific spatiotemporal configurations of social, cultural, and sexual relations of power. It is through these capacities, orientations, and habituations that we inhabit, experience, act upon, and connect with the world we live in.

2. THE LIVED BODY IS INTERTWINED WITH THE WORLD.

It is through our bodies that we are fundamentally intertwined with our environments. As Gail Weiss (1999) stated, being embodied primarily means being enmeshed in a world of nonhumans. The entanglement stems from the fact that the substance of the human body is "ultimately inseparable" from its environment (Alaimo 2010). Or, as Merleau-Ponty puts it, the body is always already "caught in the fabric of the world" as "the world is made of the same stuff as the body" (Merleau-Ponty 1964, 163; see also Merleau-Ponty 1968, 248). Thus, the body-subject is not only situated in but also made of the world—it is an integral part of the vital materialities that the world is composed of. Indeed, one can only understand the lived body, Merleau-Ponty states (1968, 250), by and through the flesh of the world it is made of.

Merleau-Ponty's relational ontology adapts the notion of intersubjectivity to the more-than-human world we live in. It thus makes it possible to investigate the role of our corporeal connection to the nonhuman world in the formation of our political agency and subjectivity. What is important here is that the notion of the body-subject provides a conceptual space to interrogate experience, political agency, and subjectivity as processes that are anchored in the lived body. This conceptual potentiality of the body-subject as intertwined with the world but not dissolved in it differentiates Merleau-Ponty–inspired feminist/critical phenomenology from the contemporary new materialist and posthumanist approaches. Even though the embodied subject and the world are always already inter-woven and intertwined, and even though we live as porous flesh in a fleshy world, we still experience the world from a particular point of view, from a particular location, through a particular (spatial, cultured, gendered, racialized) body. If we take the lived body not as an isolated, autonomous unit, but as a "prolongation of the world" (Merleau-Ponty 1968, 255), as a conceptual anchor, it becomes possible to account for the lived, embod-ied experiences of women as derived from and entwined with the material world they inhabit.

Developing such a relational, body-centered perspective helps us account for our "invisible" habitual practices within a more-than-human lifeworld and our bodily, sensory, and affective connections with the non-human objects, organisms, and entities that are established through rou-tine everyday experiences. What I propose here is similar to what James Ash calls "visceral methodologies," in the sense of using the body as a medium "to bring background or previously undetected non-human objects and forces to the forefront and so enable them to be studied and analyzed" (Ash 2017, 206). We have seen impressive examples of visceral methodologies in art, which usually prefigures conceptual thought, such as the earth-body sculptures of the Cuban artist Ana Mendieta (1948–1985), which fuse Earth—land, water, fire—and the human (female) body into performative pieces to explore various themes such as history, mem-ory, place, and violence. Using the body both as the subject and the object of the art piece, Mendieta reconfigures the Earth, that is, the nonhuman, not as separate from but as an extension of the human body (and the body as an extension of the Earth) that shapes us—our histories, memories, identities, belongings, and beliefs—in the most fundamental ways.[32]

It is the body's double existence, both as subject and object, as sensible and sentient, as a "thing among things" and as "what sees and touches them" (Merleau-Ponty 1968, 137; see also Grosz 1994; Esposito 2015; Rosa 2019) that places it ontologically and epistemologically in a unique position to unravel the "order of things" of which it is a part:

It is the body and it alone, because it is a two-dimensional being ["the sensible sentient"] that can bring us to the things themselves, which are themselves not flat beings but beings in-depth, inaccessible to a subject that could survey them from above, open to him alone that, if it be possible, would coexist with them in the same world. (Merleau-Ponty 1968, 136)

From a phenomenological point of view, focusing on the lived body's sensory experience is a useful methodological point of entry to highlight our formative encounters with nonhuman entities and environments and "to identify the traces of the more-than-human" (Clare 2019, xviii) in the formation of embodied subjectivities. In this sense, even though a critical (and post-) feminist phenomenology shares the objective of "linking human corporality with nonhuman life processes" (Oppermann 2013, 27) and with new materialist and posthumanist approaches, its way of going "back to the things themselves," as Husserl famously put it, differs. It aims to attend to the agency of things, organisms, and environments without dissolving the first-person accounts of experience altogether, using the lived body as an anchor.

SITUATING THE LIVED BODY WITHIN PATRIARCHY

The lived body situates us within a more-than-human lifeworld, which is also a world of power and domination. Radical as it is, women's political agency against HEPPs in the East Black Sea Region—which is bound to their lived experience of and bodily connection to river waters—appears contingently and contextually as they emerge within a specific configuration of power relations, social structures, and cultural values. Patriarchy stands at the intersection of these relations, structures, and values, shaping and constraining the bodily agencies of women. This is not to say that men actively deny women's agency; men and women fight against HEPPs together in most cases, even though their motivations and narratives differ. It is the structures, relations, and values maintaining the patriarchal organization of everyday life that women constantly have to negotiate and bargain with.[33] This process of negotiation is common to women's environmental struggles, especially in environmental justice movements. Women's political subjectivities in the context of environmental activism grow within the cracks of patriarchy, not against it, entailing a potential to transform it from within, at least to a certain extent (Krauss 1993, 1998; Stein 2004; Bell 2013).

The everyday realities of women as political agents are often too complex and multilayered to be represented by a particular model. Women are the beating heart of the anti-HEPP movement. Still, they are severely underrepresented in the publicly visible leadership positions within movement organizations—including the local organizations—as in many other movements in many other parts of the world (Di Chiro 1992; Seager 1993; Brown and Ferguson 1995; Sasson-Levy and Rapoport 2003; Kurtz 2007; Buckingham and Kulcur 2009). The same gendered roles and division of labor that put women in a closer, embodied, sensory relationship with their environments and provide them with a legitimate position and voice within the movement, as well as in the eyes of the broader public, also prevent them from taking up representational roles. One side of this is that women are overwhelmed with household responsibilities and do not have the time and luxury to undertake representative positions, which generally are voluntary and unpaid, come with many responsibilities, and require a serious time commitment. On the economic side of things, as most local, regional, and national anti-HEPP organizations and the movement in the broader sense refuse to accept funding in order to protect their independence, people who take representative positions mainly cover their expenses out of their own pockets. As men have more control over economic resources than women in rural Anatolia, it is easier for them to take up such positions.

Another factor is the overwhelming effect of patriarchal norms and values in rural Anatolia, including the East Black Sea Region, that place women within the family and household—which extends to the natural landscape but is still framed in a contrasting relationship to the public and political sphere. It might be threatening for many men to imagine, let alone support and respect, their wives as political figures in public, representative positions. Two incidents I experienced during an anti-HEPP festival in the village of Boğazpınar (MR) in the summer of 2014 illustrate the ambivalence of men who are against HEPP projects regarding the position of their wives within the movement. The first example is Semiha Teyze,[34] a woman who was in her seventies when she hosted me and two other young women, journalists and environmental activists from Istanbul, during the festival. During our stay, we established a personal connection, and in one of our conversations, she read us the poems she had written against HEPPs. We were very impressed with the poems and encouraged Semiha Teyze to read them on stage at the festival. She did so, and surprised us by taking the stage by storm with her confidence and poise. It turned out to be a great success, as her poems were welcomed with enthusiastic applause by the audience. Semiha Teyze was happy and proud afterward, with everyone encouraging and congratulating her— everyone except her husband, that is, who was clearly unhappy that she had been on the stage.

Another example is Güler, a young woman in her early thirties at the time. She was very active in organizing the festival, helping form a choir for children and a folk-dance group for adults. Her husband, however, did not let her participate in the folk-dance group, even though he is on the board of the local anti-HEPP organization. It is thus not surprising that the very few women in representative positions of national and local organizations are single women, such as Kamile Kaya. A photo of her in a mosque making a presentation on the socio-ecological destruction of HEPPs, standing before a male crowd without a headscarf, is what inspired me to study the movement in the first place. Kamile, an architect in her forties whom I had the privilege of meeting and conducting a recorded conversation with in her hometown of Ardanuç (EBR), was the only woman on the board of *Derelerin Kardeşliği Platformu* (Brotherhood of the Rivers Platform), which was the primary national network of the anti-HEPP movement, with connections across the country.

In his seminal book *Slow Violence and the Environmentalism of the Poor* (2011), Rob Nixon discusses Rachel Carson's and Wangari Maathai's intellectual work and environmental activism under the title "Environmental Agency and Ungovernable Women." What made them ungovernable was their independence both as scholars—they were both highly educated but not constrained by institutional affiliations—and as unmarried women. In the eyes of the scientific and political establish-ment, both of which were (and are) dominated by men, Carson and Maathai were "uncontrollable, unattached," without husbands to rein them in (Nixon 2011, 146). As in the examples of Carson and Maathai, as well as of Kamile Kaya, being unmarried seems to provide women with freedom, albeit a limited freedom in a patriarchal society, to emerge as public and intellectual figures, activists, and leaders. This is also the case for Seniye in Fındıklı (EBR) and Esin in Alakır (MR), the only other women I met who acted as local representatives and were members of the boards of their local anti-HEPP organizations. Both Seniye and Esin were university educated but had returned to their villages after graduating to live with their families and work as family farmers. Both were also single and in their mid-to-late thirties. Even though they lived with their families, they were relatively free from patriarchal pressures as single women. Nevertheless, they both still had to battle hardships to pursue their public activist roles—household responsibilities in the case of Seniye and economic difficulties in the case of Esin.

One incident in Karacaören, located in the valley of the Alakır where Esin lives, illustrates activist women's challenges. While we were talking in front of her family house, Esin asked Ahmet, the representative of the local anti-HEPP Karacaören Nature, Culture, and Tourism Association, if they could take her to the abovementioned anti-HEPP festival in Boğazpınar. The festival was going to be a meeting for anti-HEPP movements and local communities across the Mediterranean Region. Ahmet said that he and another local male representative, a local dentist, decided to go by bus instead of by private car, so she would have to pay for the ticket herself if she wanted to go with them. Esin, a subsistence farmer with very limited economic resources for her personal use, did not have money to spare for the bus ticket. She felt, as she shared with me later, sad, disappointed, and left out, as no one considered her financial situation.

All these effects of the patriarchal organization of everyday life practices, in relation to social institutions and cultural norms, shape women's environmental activism against HEPPs, as well as, and together with, their bodily, somatic experiences that connect them to river waters. As Diana Coole (2005, 135) states: "[O]n the one hand, political agents emerge from and are motivated by diffuse experiences that are lived and communicated by the body, and on the other, they emerge within and into a field of forces that incites, shapes and constrains their development while subjecting them to a transpersonal logic of collective action." It is thus essential to reiterate that bodily senses, affects, and emotions are lived by particular people in particular contexts (Tilly 2004; Clare 2019). And the body-subject, as formed through bodily experiences and encounters in a more-than-human world, is situated, spatially and historically, within specific cultural, social, and sexual relations of power.

3 Sense, Affect, Emotion

BODILY EXPERIENCES OF RIVER WATERS AND EMER-
GENT POLITICAL AGENCY

> I feel like a piece of land, parched and thirsty, when I am in
> Istanbul. As soon as I set foot in Fındıklı [district], by this
> river, my body comes to life, like when you water the land
> after a long dry season, and it absorbs it immediately.
>
> Semra, a woman in her late fifties in the village of Gürsu

Semra, who was in her late fifties when we talked in her garden in the village of Gürsu (EBR), on the banks of the Arılı River, is a retired nurse who divides her time between Istanbul and Fındıklı. Her words point to a strong affinity and affectivity between the human body (Semra's body) and bodies of water (the Arılı River) that is often overlooked in the scientific literature on human relationships with water.

There is, of course, an aspect of instrumentality in our relationship with water and water bodies. Water is the primary life-giving, life-sustaining element, and no living being can exist without regular water intake. Water bodies are utilized for transportation and logistics and for producing energy. In addition, human communities use water in the household for washing, cleaning, and personal hygiene, and in agriculture and industry to grow crops and process products. Thus, the utility of water is a key reason why human communities have been settling around water bodies for thousands of years. Our fascination with water bodies, however, is not limited to utility. Utilitarian framings of water fall short of explaining our "fundamental," "spontaneous," and "existential" attraction (Stefanovic 2020b) to water. To illustrate that "ontological draw," Stefanovic cites Nichols (2014), who writes about a 2010 study conducted at Plymouth

University. Subjects were asked to rate more than one hundred pictures of natural and urban environments. They rated virtually any image containing water more highly than others. We know this appeal from our own lives, as we associate recreation, relaxation, and rejuvenation with water—think of holidays on the coast or by a lake or river, spas and thermal springs, or the desire for a room with a sea, lake, or river view. We love putting our bodies into the water, watching the river's flow, listening to the sound of the waves, and drinking from a spring.

Such experiences of water are rarely discussed in a scholarly manner, however (Stefanovic 2002a; see also Neimanis 2017; Jewett and Garavan 2019 Clare 2019), as they either become invisible, lost in the routines of everyday life, or are seen as less important than the problems, policies, and practices related to water as a "resource." Water access and sustainability are indeed major issues of our times. Consequently, water governance, management, and development are receiving more and more scholarly attention, and rightly so. The conceptualization of water only as a vital resource, though, misrepresents our connection to water, which goes far beyond its instrumental use. In this chapter, I aim to unveil the intimate corporeal connection between human bodies and bodies of water, using Merleau-Ponty's work and critical and feminist phenomenological perspectives built on his work. Beyond exploring this sensory and affective connection, this chapter builds on the empirical case of women's activism against HEPPs in the East Black Sea Region to demonstrate how bodily senses and affects condition our subjectivities and our emotional and political agencies and their limitations. To do so, I draw on phenomenology in close dialogue with water studies, especially with the anthropology of water, affect and emotion studies, and feminist literature on agency and subjectivity.

WATER, WATERS, WATER BODIES: A PHENOMENOLOGICAL APPROACH

Water is probably the most symbolically loaded element on Earth. The incredibly rich representations and meanings water entails stem from the fact that our bodies are hydrophilic, and so is our world. As Astrida

Neimanis (2017, 65) states, we are not only "bodies of water in a watery world," we are also "deeply imbricated in the intricate movements of water that create and sustain life on our planet." Water permeates and moves through organic and inorganic substances, human and nonhuman bodies, and their environments. We all are part and parcel of the hydraulic cycle—the movements of planetary waters that flow through bodies, organisms, and ecosystems. In that sense, "water defines us as embodied creatures" (Stefanovic 2020a, 3), or, in other words, "our experience of embodiment is deeply connected with water" (Strang 2005, 99). We do not dissolve in planetary waters, however; we remain distinct bodies, even though our skin is porous and we are constantly absorbing water and leaking our various water-based secretions into the world. And we encounter and experience planetary bodies of water with our watery bodies every day.

The relation between embodiment, experience, and water is multilayered: not only is our experience of embodiment watery—as Stefanovic and Strang state—but our experiences of water are also inherently embodied (Sultana 2009). We both embody water and have a strong embodied relationship with water. Our encounters with water bodies have the potential to generate the full spectrum of sensory experiences (touch, sight, hearing, smell, and taste) and affective responses that the human body is capable of. Encounters of different lived (human) bodies and bodies of water, though, generate different sensations and affects, going through similar "sensory surfaces" (Highmore 2010, 121) of our bodies. To put it differently, the sensory and affective (perceptual) capacities of our anonymous, physical bodies are realized differently in the context of specific encounters between the lived bodies and bodies of water.[1] Here I focus on the encounters of East Black Sea women with the streams of the region to explore their particular sensory-affective connection, which shapes their political agencies as they struggle against HEPPs.

We know why understanding bodies in specificity and difference is important from a feminist perspective (see the discussion on the "lived body" in the previous chapter). What is less stressed is the importance of attending to the specificity and difference of water bodies, conceiving them not as generic, but as lived entities with specific material properties. Water is represented, especially within the scientific discourse, as a homogenous, colorless, tasteless, shapeless substance (Neimanis 2017;

Walsh 2018; Stefanovic 2020b). Casey Walsh (2018, 7) defines this scientific-utilitarian representation of water as the domination of waters by water: "Rather than devote energy to understanding particular waters and how they shape diverse human ecologies, scholarship on water in the twentieth century usually treated water as an inert, universal backdrop to the question of how humans organize themselves socially and politically to utilize the substance." Our everyday experiences of water, and especially of water bodies, on the other hand, are shaped by encounters of particular human bodies and particular waters in specific spatiotemporal contexts. The material properties of these particular waters are thus constitutive of our experience of them.

There are obvious differences between the various types of water bodies—the sea and the ocean clearly differ from the lake and the pond and from the river and the creek—and our ways of experiencing them differ as well. When we talk about rivers and streams, for example, we are essentially talking about waters that flow, from one place to another, from their sources to their mouths. Heidegger famously stated that rivers determine the dwelling place of human beings upon the earth (Heidegger 1996; see also Donohoe 2020). Not only are large cities often built along rivers and streams, but smaller towns and villages are as well. Maybe less known is Heidegger's definition of rivers as both a place and a journey (Heidegger 1996). Rivers are "local at all points" (Bromber, de la Croix, and Lange 2014),[2] but they are also in constant motion. Yet not all rivers give the same impression of place or movement. Their size, volume, and accessibility; the slope of their streambeds; their clarity and transparency; the taste and texture of their waters; the landscapes within which they flow—all of these characteristics shape our perception of rivers. Hence, a phenomenological approach that focuses on our lived experiences of river waters should focus on "local river systems" and their "specific instances" of bodily interactions (Deane-Drummond 2018).

Accordingly, this chapter focuses on the local river systems of the coastal part of the East Black Sea Region, where the rivers are small in size and volume (more like streams [*dere*] than rivers), cascading down naturally steep slopes leading from the Kaçkar Mountains to the Black Sea through high, densely forested valleys along which villages are located. Small natural pools form in various spots and serve as little swimming pools for

Figure 1. The Arılı River (Fındıklı District, Rize Province). Photo by the author.

children and adults. Their waters are cool, clean, and crystal clear; they "shimmer and flicker with constant movement" (Strang 2015). It is primarily, though not exclusively, women who live in intimate bodily connection with river waters, as they are "intimately interlocked" (Murton 2012, 90) with their immediate environments by the gendered organization of everyday life and division of labor.[3] Rivers are the central elements of the landscape in the region, and as the houses are dispersed along them, living within this particular landscape in a particular place (village) translates into living by river waters. Hence the experiences of the river waters, and the sensory-affective responses they generate, are "mundane" for East Black Sea women, being an integral part of the routine, habitual patterns of everyday life.

As phenomenologists know very well, the lived experiences of everyday environments, "taken-for-granted ways of being in and knowing about the world" (Stefanovic 2020a, 4), often get "lost in translation," flattened out or rendered invisible by "scientific" accounts of action and interaction.

Figure 2. The Cihani River (Arhavi District, Artvin Province).
Photo by the author.

This is why Husserl developed the method of phenomenological reduction: bracketing our positive knowledge to return to the description of our
lived experience (Landes 2012). As much as we know—first and foremost,
thanks to Merleau-Ponty (2012, lxxxv)[4]—that complete reduction is not
possible, a methodological move is required to return to the lived experience that has been rendered invisible. Husserl articulated this bracketing
as a conscious move of the transcendent subject, but I argue that it can
just as well be an effect of a sudden change that happens not in the subject's consciousness, but in her phenomenal world.

Dramatic and unexpected events that result in sudden dispossession of the routinely experienced environments, what social movements scholars would define as "suddenly imposed grievances" (Walsh 1981), can engender such an effect. Suddenly the routine becomes visible; the mundane becomes special. The prospect of loss, thus, acts as a vehicle of phenomenological reduction that brings us "back to the things themselves," as Husserl once put it, particularly to the thing we are about to lose—in this case, the river. The prospect of losing the river waters has caused East Black Sea women to recognize, value, and voice their everyday, embodied experiences. When we understand the "suddenly imposed grievances" that the HEPP projects represent as violent events that provoke "bracketing," it becomes clear why the East Black Sea women begin to speak so passionately about their embodied sensory connection to rivers, which is otherwise a routine, banal experience that is invisible under normal circumstances.

MULTISENSORY EXPERIENCES OF RIVER WATERS IN THE EAST BLACK SEA REGION

If our body is "our manner of being at the world" (Olkowski 2006, 3), our senses are the "modalities" of our bodily existence (Tilly 2004, 15). They are the media of our bodily interaction with the world; we perceive, experience, and reach out to the world with our senses. And it is only through the sensory perception that the world appears to us and "reveals itself" (Le Breton 2006, 9). This dual nature of the relationship between the self and the world is important here, as the sensing self is always already a part of the sensible world. Thus, perception is never an act of the sensing self alone; it is instead "a communication or a communion," "a coupling of our body with the things" (Merleau-Ponty 2012, 334). Hence, one perceives "with both his body and his world" (Merleau-Ponty 2012, 216). Or, as neatly formulated by Helen Fielding (2017, ix), "sensing is sedimented through engagement with the sensible" (see also Abram 2017).

Seeing, for example, involves not only the seer but also the visible, of which the seer is a part. Seeing, thus, is not a subject-centered activity, but emerges out of the communication of the visible with itself. In other

words, "I see with a seeing that is not mine alone" (Butler 2015, 60). This
is what Merleau-Ponty implies when he talks about the coiling over of the
visible upon the seeing body and the tangible upon the touching body
(Merleau-Ponty 1968, 146–7). From a phenomenological perspective,
sensory perception indexes our relation to and relationality with the more-
than-human world (see, e.g., Clare 2019). East Black Sea women's narratives about their sensory and affective connections with river waters
establish sensation as a communication or a communion with rivers,
manifested in a multisensory experience. Different from the oceans and
the seas, the experience of which involves olfaction, fresh waters, at least
when not contaminated, do not strike us with their smell. Sight, touch,
sound, and taste, however, appear to be the primary ways in which we
commune with river waters.

The sight of the flowing river has calming, relaxing, hypnotic, and mes-
merizing effects. Anthropologists of water, such as Veronica Strang, write
about these effects: "Any lake, pond or river is a visual magnet, drawing
people to sit gazing at the water, mesmerized by its glittering, dancing
lights" (Strang 2015, 49). Seeing a river's cascading rush or calm flow is a
significant aesthetic and affective experience, as expressed by Elif, a
woman in her thirties who showed me the Arılı River (EBR) and said,
"Look how beautiful it is! They want to take this beauty away from us.
They want to leave us with the dry rocks of the riverbed." Rivers are not
only visual magnets that engender relaxation and contemplation; they are
also public places, places of recreation and entertainment, and places of
shared beauty and joy. As we were talking within a crowded group of family members in front of her house in the village of Yaylacılar, also located
in the Arılı Valley (EBR), Refiye, a woman in her sixties, pointed at the
river: "We live here, in this narrow valley, only with the joy of the river
[*derenin neşesi*]. When it is gone, it means we should also go." What she
expresses with the phrase "the joy of the river" is the aesthetic and affective experience of looking at flowing water. The joy she mentions is the joy
of living in that particular physical and affective landscape, which is ani-
mated by fast-flowing, cascading waters.

Ülker, a middle-aged woman from the village of Aslandere (EBR),
described what Refiye articulated as joy a bit differently as we were
having tea after dinner in the house of her neighbor Seniye: "Rivers are

our celebrations [*şenlik*]. I went by the [Çağlayan] river today before I prayed. I just sat by it and watched it. It was so beautiful, greenish blue. I watched the fish swimming in it. I stayed there for some time and returned to pray." Ülker's description demonstrates the centrality of the mere sight of the river for the local community and reveals the multiple affective registers that a glimpse of the river can engender. She felt the urge to go and watch the river before her prayers, and it seems to have had a calming and grounding effect on her. But she also talks about the beauty of the river, the colors, and the fish. And the beauty and vitality radiated by the river's flow make it a "celebration"—of life itself. In a video reportage by BBC Turkish, Ayşe Kurdoğlu, a middle-aged woman, also stresses the importance of gazing at the river:[5] "We live here [in the village of Ulukent—EBR] with the river. We look at the river every day. We cannot live without it." Associating the sight of the river with some sort of companionship (we live here *with* the river—with the *sight* of it) recalls Merleau-Ponty's analysis of perception as a communication or a communion with the perceived. In this case, beyond the joy that the sight of the river generates in the human inhabitants of the valley, seeing it every day manifests a relationship—of living together, of coexistence and companionship.

As our experiences of water bodies and our everyday experiences in a broader sense are multisensory, the sight of freshwater streams is usually accompanied by sound, especially if they are fast-flowing, steep mountain streams as in the East Black Sea Region. The constant sound of flowing water could be disturbing to someone who is not accustomed to it, but it is the sound of life for people who have grown up and lived all their lives by a stream. Cemal Can is an older man from Loç Valley in the West Black Sea Region, which shares many geographical features with the East Black Sea valleys and is similarly threatened by HEPP projects. Speaking to the camera in the documentary film *Sudaki Suretler* ("Figures in the Water"), Can says, "When I get up, I first look at the [Devrekani] stream. It makes a sound similar to the wind. When I see it, I feel at ease, relieved."

The sound of the river is commonly identified with life, peace, and memories of the past. Nadire Yıldız, an eighty-year-old woman from the village of Konaklı (EBR), tells a reporter from the newspaper *Evrensel* that she cannot sleep without the burbling of the river, which is like a lullaby to her (Uğurlu 2015). Equating a river's sound to a lullaby evokes the calming, soothing properties of water sounds, be they of ocean waves, rainfall, fountains and springs, or, in our case, streams. It is not a coincidence that sounds of water, especially in the form of fountains, are traditionally used to soothe mental health patients (Strang 2004). The soothing properties of flowing, falling, or lapping water are utilized nowadays by commercial apps developed to assist mindfulness, relaxation, and sleep. For Nadire, however, this "lullaby effect" goes far beyond the instrumentality of listening to water sounds on smartphones to fall asleep. The river's sound is the sound of life for her, to which she wakes up and with which she falls asleep every day. Constantly hearing its sound calls forth a feeling of togetherness, a companionship that makes the absence of that sound unthinkable. This is why Nuran, a woman in her sixties from Aslandere (EBR), likened the absence of the river's burble to death during our interview at her house.

Nuran's association of the river's sound with life and its absence with death is not merely symbolic. Bozok, Bozok, and Akbaş (2016) tell the touching story of ninety-six-year-old Emine from a remote village in the District of Borçka (EBR). Emine, who lives alone on the steep slope of a deep valley, told them she could not hear anymore. During their talk, however, she heard them quite well. When they asked why she thought she could not hear, she responded, "I thought I must have gone deaf because I cannot hear the river." The river was the sound of Emine's life, and she could only explain its absence by assuming she had lost her sense of hearing—hence death of the senses, if not the death of the self. In fact, what was gone was not her sense of hearing but the river itself (it had been rerouted via pipes to a HEPP tributary), and with it, the background sound of a long life. This story demonstrates how the river mediates the relationship between the embodied subject and its environment by stimulating bodily senses and affects. Its perpetual existence shapes perception in such a way that it affects how the embodied subject experiences her sensual capacities.

Another aspect of this relationship established between a river's sound and life itself is expressed by Seniye, a young woman and a well-known local activist from Fındıklı (EBR), in a video reportage by BBC Turkish:[6]

"You know that people all eat to stay healthy. Well, we listen to the sound of this river [the Çağlayan River flowing behind her as she talks] to stay healthy. We get stronger as we hear it. We will get weak if it goes silent." This association of a river's sound with vitality and health is another common motif, articulated by several of the women with whom I spoke. In another, earlier BBC Turkish report on anti-HEPP movements, Neşe Terzi, who is also from Fındıklı, from the village of Meyvalı, states: "It is enough to sit by this beautiful [Arılı] river, on its rocks. How nice is its sound. . . . There is no need to go to the doctor" (Kasapoğlu 2013). Neşe refers to the therapeutic, healing properties of water—flowing water and its sound, in this case—which have long been recognized and used in many different parts of the world (see, e.g., Strang 2015; Walsh 2018).

The therapeutic properties of a river's sound are related not only to its calming, soothing, relaxing impact. Those grounding properties go hand in hand with more stimulating ones, as the burbling, splashing fresh water also animates, invigorates, and revitalizes. Years before she spoke to BBC Turkish, Seniye had told me about such properties of the river's sound as she was guiding me around her village of Aslandere (EBR) when I visited in the summer of 2013:

> Yesterday I was working in our hazelnut field close to the river. I was tired, so I went to the river, put my [basket of] hazelnuts on a rock, and slept with the sound of the river for about half an hour. I woke up so refreshed and energetic, and I started to work again. . . . The river is life and soul [*hayattır ve candır*] for us. I would probably feel utterly empty if I did not hear the sound of the river.

Her use of the word *empty* is worth mentioning here, as it indicates a conception of the self in which the "internal" state is conditioned on its immediate interaction with the "outside." This differs from the classical conception of the solid human body, the outer borders of which keep its contents firmly inside, independent of external factors. A relational articulation of interiority and exteriority, resting on a "continuous transgression" of the body into things and things into the body, is a distinctive feature of Merleau-Ponty's approach to sensory experience (Lefort 2012; see also Coole 2007).

AFFECT, *EINFÜHLUNG*, AND THE FLESH: MORE-THAN-
HUMAN ENCOUNTERS BEYOND HUMANIST ONTOLOGIES

Even though Merleau-Ponty has often been counted within the humanist
tradition due to his close ties to French existentialism, many critical and
feminist scholars of phenomenology have pointed out his later work, especially his concept of flesh, to counter his misrepresentation as a transcen-
dental humanist (see, e.g., Lingis 1968; Barbaras 2001; Coole 2007;
Fielding 2017). The premise of his (anti-humanist) relational ontology,
however, is apparent even in *Phenomenology of Perception,* as his corporeal take on perception counters the previously established idea of sensory
experience as purely subjective. In his account, the corporeality constitutive of perceptive experience is not limited to the body of the sensing subject, as discussed earlier. Perception is rather an "active interplay," an
"ongoing interchange between my body and the entities that surround it"
(Abram 2017, 55–57). This understanding of sensory perception as communication or coupling makes it possible to think of senses in their close
connection to bodily affects. It is also the starting point for Merleau-Ponty,
as he states in the very first sentence of the chapter on sensation in
Phenomenology of Perception (2012, 25) that sensation might first be
understood "to be the manner in which I am affected."

Affect theory scholars relate affect, similar to Merleau-Ponty's understanding of sensory perception, to bodily capacities (to affect and to be
affected) and to the body's ongoing immersion in a more-than-human
world of encounters. Affect arises out of bodily encounters through
sensate—haptic, aural, gustatory, olfactory, and visual—perception and
refers to the vital forces and intensities that are produced within the visceral world of encounters "that find corporeal expression in the feeling
body" (Ash and Simpson 2016, 55). Affects, thus, are not produced but
felt by the lived body, "which widens, tightens, weakens, trembles, shakes,
etc., in correspondence to the affects and atmospheres we experience"
(Fuchs 2013, 613). Even though it is challenging to pin down affect meth-
odologically, given affect theorists' suspicion of personal accounts of affective experiences (see, e.g., Anderson and Harrison 2006; Pile 2010), the
impressions human and nonhuman others leave on the body (Ahmed
2004) are a good starting point. One of those bodily impressions is the

feeling of the vitality of the more-than-human life as one's own vitality and aliveness (see, e.g., Clare 2019).

East Black Sea residents describe "the dynamism of felt intensities" (Ash and Simpson 2016, 55) that passes through their bodies during and after their corporeal, sensory interactions with river waters. Hamsici (2010, 172) cites Naşide Şimşek, a woman from Meydancık (EBR), who talks about their local river having such an impact of vitality, aliveness, and "coming to life" on her mother. Although she is an elderly woman who recently had serious bypass surgery, she starts to move "like a flea" as soon as she comes to her village by the river. "We come here for the river. Without the river, this place would lose its meaning," she says. For Naşide, the river waters animate the nonhuman landscape and the human bodies it flows within and through. The animation of her mother's old and sick body illustrates how bodily affect, as a product of more than human encounters, acts as a visceral force to drive us toward movement (Gregg and Seigworth 2010). Understanding sense and affect from a similar relational perspective—that is, as experiences that emerge within bodily interaction and encounters—it becomes possible to use them together to analyze concrete bodily encounters, as most bodily encounters evoke sensory perception and affective register simultaneously. In other words, "senses and affect bleed into each other" (Highmore 2010, 120), as in the case of our bodily relationship with flowing water.

Conceptualizing sensory perception as communication or as a coupling, similar to an affective register, provides a conceptual ground on which the East Black Sea women's sensory-affective connection to river waters can be analyzed not as a purely subjective experience that is contained within the individual self, but as an intimate bodily encounter with a nonhuman entity. When one frames sensory perception as communication or as a coupling, which is the context in which bodily affects are engendered, the narratives of symbolic identification or merging with nonhuman entities, such as rivers, become meaningful—above and beyond any cosmological or spiritual significance they may have—as manifestations of an intimate sensory connection. Elif, whom I mentioned above, signifies such a connection when she says, "We are all connected to the river, older women even more so. They have lived all their lives with the waters of the river. They have *merged* with it." The lifelong sensory, tactile relationship with river waters enacts some sort of coupling between women's bodies and river waters, which is communicated as a merging. The association of rivers with the veins of the body, a common trope used by local communities, could be seen as another manifestation of a communion engendered through sensory connection. In this trope, the term *can damarı* ("lifeblood") is often employed to stress the vitality of the river waters for community life.

Ayşe Cengiz, a middle-aged woman from Fındıklı (EBR), articulates this common analogy for a newspaper reporter: "What they do not understand is that the river is our lifeblood. Taking our river from us is cutting off our lifeblood. That is it" (*Radikal* 2015). This idea of the river as lifeblood also extends to other vital functions of the human body, and even to life itself. Semra, whom I cited at the beginning of the chapter, expressed it poetically during our conversation: "River means human, river means life (*can*), river means breath, river means *our* blood. What more can I say?" These statements remind us of Merleau-Ponty's concept of *Einfühlung*,[7] which can roughly be translated as empathy. It is through corporeal *Einfühlung* that "the world and others become our flesh" (Merleau-Ponty 2003, 211). Merleau-Ponty claims that it is the body "as corporeal schema" that furnishes us with *Einfühlung*. The body is in a circuit with the world, intimately and emphatically relating with other human and nonhuman bodies (Merleau-Ponty 2003, 209). Consequently, the body, "as the power of the *Einfühlung*," is the source of identification, projection, introjection, libido, and desire (Merleau-Ponty 2003).

In the typical descriptions of the ecological damage caused by HEPP construction, these processes of identification, projection, and introjection manifest themselves as pain felt in the lived body. Leyla Aydın, a fifty-seven-year-old woman from Senoz Valley (EBR), for instance, says:

> Plants have died, bees are gone. . . . They tore our mountain apart. . . .
> Whenever I look at the mountain, I feel like it is crying. . . . I feel the pain in
> my heart. Taking our land means taking our lives. . . . In our villages, we are
> nature itself. We are one with our land.

Aysegül, a woman in her thirties from the village of Konaklı (EBR), used the metaphor of a knife to describe the physical pain she feels as she watches the excavators digging up the riverbed in her village. "Here," she said, putting her hand in the middle of her chest as we talked, "I feel as if they are stabbing me right here." Meryem, who was in her fifties when we talked on the terrace of her small guest house in the Kamilet Valley (EBR), used the same analogy: "I feel like a sharp knife enters my chest when I hear the sound of the excavators working." Leyla, Aysegül, and Meryem's words counter the conventional understanding of the human individual as an independent entity, strictly separate from her natural environment.

The concept of *Einfühlung*, which helps deepen the phenomenological analysis by clarifying the narratives of corporeal identification or merging, belongs to a broader relational ontological framework introduced in Merleau-Ponty's later work. Merleau-Ponty articulated this ontological framework, the so-called ontology of flesh, in *The Visible and the Invisible* (1968). The concept of the flesh refers to the vast network of Being, the common web of bodies, things, organisms, and ecosystems shared by persons and things, in which tactile and sensory relations occur. Our sensory and affective connections with certain elements of the nonhuman world are enabled by the fact that our lived bodies are embedded and enmeshed in this vast network of more than human bodies, energies, and materialities. Flesh, then, could be defined as the very condition of tactility, of touching and being touched, of seeing and being seen, and their inherent reversibility, "out of which both subject and object, in their mutual interactions, develop" (Grosz 1994, 103; see also Merleau-Ponty 1968; Evans and Lawlor 2000; Al-Saji 2001; Butler 2005; Olkowski 2006; Stawarska 2006; Coole 2007; Simonsen 2012; Fielding 2017).

The word *flesh* indicates a common materiality shared by the body-subject and the world. And it is the very materiality of the body, its flesh, that opens us to the flesh of the world, the corporeality of other things and other bodies, putting us in a circuit with them: "The body as touching-touched, seeing-seen, the place of a kind of reflection, and, thereby, the capacity to relate itself to something other than its own mass, to close its circuit . . . on a sensible exterior" (Merleau-Ponty 2003, 209). It is through the flesh of our body that we become a part of the flesh of the world. In other words, it is through our sentient bodies and their connection with a sensible world—a world that touches us as we touch it—that we can feel the power of *Einfühlung*, that is, we feel as one, as "the world and others become our flesh." The lived body thus becomes a "prolongation of the world" (Merleau-Ponty 1968, 255), "a passage to an outside" through which we "incorporate" that outside. What Merleau-Ponty conceptualizes as *Einfühlung*—the bodily process through which the narrowly defined borders of the individual self are transcended and the lived body "flows over into the world" (Simonsen and Koefoed 2020, 22)—is also articulated by ecologists, environmental anthropologists, and relational materialist theorists as the expanded, expansive, and ecological self (Fox 1986, 1990; Naess 1989; Gaard 1993; Ingold 2000; Milton 2002; Bennett 2010; Braidotti 2013; Descola 2013). This feeling of expansion is engendered within an intercorporeal interworld of human bodies with nonhuman natures, displaying the capacity of the body to extend into the world along sensory pathways (see, e.g., Strang 2005; Bennett 2010).

New materialist feminist theorist Stacy Alaimo explores this feeling of expansion in *Undomesticated Ground*, drawing on early twentieth-century novels written by American women. Her analysis of *Call Home the Heart* (by Fielding Burke) and its main character Ishma scrutinizes the "expanding, spatial self," that is, "self becoming boundless when it disperses into the expansive realm of nature" (Alaimo 2000, 115). What is interesting in the case of Ishma, a woman struggling with poverty and the hardship of rural life, is that she experiences nature not as what it is imagined to be in the conventional feminist wisdom, that is, as "a repository for cultural moralism and prejudice" (Alaimo 2000, 19), an adamantine and eternal system of things that imprisons women in a certain—domestic and inferior—status. On the contrary, nature appears to be an oppositional, emancipatory space that provides women with much-needed freedom and pleasure outside the domestic field. The feeling of expansion, of "flowing over into the world," as we see also in the narratives of East Black Sea women, is a facet of such freedom, the freedom of transcending the individual self—a self defined by social constraints.

TOUCHING THE WATER, TASTING THE WATER: AFFECTIVITY, REVERSIBILITY, AND THE LIVED BODY

Skin, as our largest organ, "proves to be the critical interface in all of our bodily relationships with the world" (Rosa 2019, 48; see also Ahmed and Stacey 2001):

> We develop not only our sense of the world—as what resists us, as what we encounter out there—but also our sense of self, of what is peculiar to us, and of the boundary or distinction between the two through tactile experiences, i.e., by touching, grasping and handling as well as by being handled, grasped and touched. . . . Self and world encounter each other constantly and directly, if under constantly changing conditions, at and through the skin. (Rosa 2019, 48)

That is to say, it is not only that a person becomes a part of the flesh of the world through the flesh of the body, but she also senses the flesh of her body only within the flesh of the world. If I know myself through my relation to things (Merleau-Ponty 2012; see also Coole 2007), to human and nonhuman bodies and environments, tactility is the main form of this relation through which the "I" appears (Butler 2005).

Feminist critics of Merleau-Ponty rightly point out his prioritization of the gaze as the archetype of the reversibility of the sensible, as in his last published essay, "Eye and Mind" (published in English in 1964 in *The Primacy of Perception;* see, also, Irigaray 1993; Butler 2005). They argue that relations of tactility provide us with conceptually and analytically more effective ways of exploring both the idea of the reversibility of the flesh and the idea of embodied subjectivity as it appears within the material, intercorporeal flesh of the world. This dual, reversible sensation of touching and being touched is powerfully manifested in our bodily relations with water, especially in the case of our immersion in water bodies, such as ponds, lakes, oceans, seas, and, of course, rivers and streams. Swimming and being immersed in water, in general, has strong affective power on our bodies as it fosters an unusual corporeal experience of floating, of being weightless or feeling light (see, e.g., Drew 2017). In this sense, being in water might be as close as we can get to flying. This strong affectivity results from the "undoing of the terrestrial body," in the words of Stephanie Clare (2019, 23), which involves "remaking and reworking of muscles and organs" in our bodies. There is, of course, the changing effect of gravity, as it is coupled with buoyancy, but how our bodies interact with these physical forces in water has to do with the material composition of our bodies as well. And as we immerse our bodies in the water, we do it as watery bodies, as bodies made of "wet matter" (Neimanis 2017).

Swimming in water is thus encountering an outside that is also an inside (Weber 2016; see also Bachelard 1983). Even though Merleau-Ponty's concept of the flesh refers mainly to an ontological realm of being and should not be reduced to one particular matter or substance, one cannot help but think of water as a shared web of materiality that folds in on itself to create different forms of life. Immersion in water also produces a rich range of sensory, affective, and cognitive responses, from feelings of well-being to relaxation and heightened imaginary activity (Strang 2005). Bodily interactions with bodies of fresh water, like rivers, can be experienced as soothing, calming, and uplifting, as Garhwali women describe them (Drew 2017). As a textbook example of bodily affect, the encounter of fresh, wild waters with human bodies produces strong bodily sensations, translating into vitality, aliveness, and joy. Through the touch of fresh water, we feel all parts of our bodies anew, leading to shivering and goose bumps on the skin and an increase in blood circulation and heart rate. Photographer Liz Seabrook, who conducted a project with female wild-water swimmers, registers the transformative affectivity that submersion in wild, fresh, cool waters engenders. Her pre- and postswim photos of the famous wild-water swimmer Kate Rew record a discernible change: "[B]efore there is a slight unease, but afterwards she's relaxed yet stoic—there is an openness that wasn't there before" (Seabrook 2020, 144).

Even though the relationships between the bodies of water and the bodies of women that Seabrook captures in her photographs are established within a cultural and spatial context that is completely different from the East Black Sea context that I am investigating—the women photographed by Seabrook are all white and middle class, from Britain—one can observe similar sensory experiences and affective responses—opening, soothing, calming, uplifting—among British wild-water swimmers and the East Black Sea women immersing themselves in their local rivers and, for that matter, among the Garhwali women bathing in the Ganga (Ganges) studied by Georgina Drew. It is common among East Black Sea women to go into the river fully clothed, especially after working in the tea and hazelnut fields in the summer, to relax and refresh. They often mention the animating and cooling affectivity of the river's touch. Ayşe Cengiz, for example, is quoted in a newspaper article saying, "The river flows just in front of my house. Whenever we swelter, working in the tea fields or at home, we go in the river to cool off" (*Radikal* 2005). The cooling touch of the river seems to be a part of the bodily regulation regime for women and of the social reproductive processes in the region, as the cooling-reviving affective encounters of women's bodies and river waters help women to acquire the physical and emotional strength to keep working in the fields and at home.

The river's touch, though, is not purely instrumental or instrumentalized. Immersing oneself in the flowing water and swimming in the pools that form in the river's bends are important forms of recreation and bodily activities that generate joy and happiness. And East Black Sea women are particularly fond of this activity. Gülşen, for instance, a woman from Aslandere (EBR), described herself as a frog to me to express her love of swimming in the river, then she continued: "We do not have beaches or swimsuits. We swim here in our clothes. . . . We have such a beautiful and clean river that you can drink from it while swimming. . . . We cannot live without the river. *I cannot* live without the river." This ostensibly simple comment in fact contains a great deal of richness. In the first two sentences, she differentiates swimming in the river from swimming in the sea, as the whole habitus of the latter is relatively foreign to her, whereas freshwater swimming is accessible (as the river flows within the village) and is culturally acceptable, as women swim there with their clothes on (wearing swimsuits, however, is not taboo, and children and young women do sometimes wear swimsuits).

The second point is the simultaneity of haptic and gustatory senses: drinking from the water you are swimming in, which is not something most people would normally want to do given the water pollution affecting fresh waters as well as seas and oceans. The sense of gustation contributes to the overall bodily affect, as the taste of the river is experienced as cooling, reviving, and vitalizing. Elif, whom I mentioned before, describes the experience: "When the weather is so hot in summer, we go and drink from the river; it is so refreshing to drink from it." Merleau-Ponty relies on the relationship between tangible and visible to emphasize sensory convergence or "crisscrossing" between different senses: "There is encroachment, infringement, not only between the touched and the touching but also between the tangible and visible. . . . Since the same body sees and touches, visible and tangible belong to the same world" (Merleau-Ponty 1968, 133–4).

The "encroachment" between touch and taste, however, stands for another aspect of the essential multi- and inter-sensoriality (Howes 2005) of our everyday experiences. Skin and mouth are two organs through which we incorporate the world and extend into it, and water is probably the quintessential subject/object of this constant process of incorporation. Beyond the well-known fact that our bodies are constituted mostly of water, our corporeal extension into the world is manifested in watery forms—we leak into the world in the form of sweat, tears, and urine; and the world interpenetrates us as our skins absorb water and as we drink. Thus, our corporeal interconnection with water illustrates what Stacy Alaimo has dubbed "transcorporeality," whereby "the substance of the human is ultimately inseparable from the environment" (Alaimo 2010, 2).

There is yet a third aspect that I would like to focus on. It is how Gülşen utters the same sentence with two different subjects—*we* and *I*—at the end of the quotation. The way she mentions "I" after "we" emphasizes her personal connection with the river from the viewpoint of the preconditions of corporeal living, something always experienced through the individual lived body. This emphasis on "I" is thus essential to delineate the embodied subject as a unit of experience. This is what differentiates a critical and feminist phenomenological perspective, which draws on Merleau-Ponty's work, from posthumanist and new materialist approaches that build on the works of Giles Deleuze, Bruno Latour, Donna Haraway, and Karen Barad. Merleau-Ponty's concept of flesh could be read in parallel with the theme of common materiality or corporeal continuity between human and nonhuman bodies, organisms, and things that had been a milestone of new materialist and posthumanist approaches (see, e.g., Alaimo and Hekman 2008; Alaimo 2010; Bennett 2010; Coole and Frost 2010; Braidotti 2013; Neimanis 2017).

Indeed, Merleau-Ponty himself famously claims that there is "an indivision . . . of my body and the world" (Merleau-Ponty 2003, 279), and that "the world is made of the same stuff as the body" (Merleau-Ponty 1964, 163): "The body, as a structure of an ensemble, that is, as an opening to the things and to others, that is, as sensing itself in the things and in others—undivided in an undivided world" (Merleau-Ponty 2003, 281).

But even though Merleau-Ponty's ontological perspective in his later work comes close to the anti- or posthumanist ontologies established by those figures, the idea of the body-subject, as an entity that has experiences, remains. And as long as common materiality does not explain how we relate to the world per se, the phenomenological ontology of Merleau-Ponty differs from new materialist/posthumanist accounts.

Corporeal continuity is a condition of sentience, as Merleau-Ponty explores with his famous conception of reversibility—of the positions of subject and object, of perceiver and perceived, of touching and touched. The body is capable of seeing and touching only because it is itself visible and tangible; we can only sense as we are surrounded by and embedded in the sensible—in the flesh. As Merleau-Ponty (1968) stresses in the famous chapter of *The Visible and the Invisible* titled "The Intertwining—The Chiasm," perception becomes possible only because the body is a variant of the flesh, and part and parcel of the sensible that it senses:

> My hand takes its place among the things it touches, is in a sense one of them, opens finally upon a tangible being of which it is also a part. . . . He who sees cannot possess the visible unless he is possessed by it, unless he is of it . . . he is one of the visibles, capable, by a singular reversal, of seeing them—he who is one of them. . . . It is that the thickness of flesh between the seer and the thing is constitutive for the thing of its visibility as for the seer of his corporeality . . . it is their means of communication. . . . If it touches them and sees them, this is only because, being of their family, itself visible and tangible, it uses its own being as a means to participate in theirs, because each of the two beings is an archetype for the other, because the body belongs to the order of things as the world is universal flesh. (Merleau-Ponty 1968, 133, 134–5, 137)

However, this common materiality—the common web of flesh—does not completely flatten or dispose the positions (of touching and touched, seeing and seen), even though they are inherently reversible. This is exactly Butler's point when she states that Merleau-Ponty's theory of reversibility overcomes the ontological distinction of the subject and the object, but not by collapsing it into a flat ontology: "The distinction between active and passive is confounded . . . without being negated in the name of sameness. . . . [T]he acts of seeing and of being seen, of touching and of being touched, recoil upon one another, imply one another, become chiasmically related to one another" (Butler 2015, 51). What is important here is that, in Merleau-Ponty's ontology, the positions of subject and object are reversible as singularities produced within an intercorporeal interworld, within the flesh "as chiasmic becoming" (Coole 2007, 171).

Yet reversibility does not mean coincidence or overlapping. As in his famous example of two hands touching each other, both hands can take both positions (touching and touched, subject and object, sensible and sentient), but not simultaneously:

> My left hand is always on the verge of touching my right hand touching the things, but I never reach coincidence; the coincidence eclipses at the moment of realization, and one of two things always occurs: either my right hand really passes over to the rank of touched, but then its hold on the world is interrupted, or it retains its hold on the world, but then I do not really touch it—my right hand touching, I palpate with my left hand only its outer covering. (Merleau-Ponty 1968, 147–8)

There is thus "a shift, a spread between them" (Merleau-Ponty 1968), "a slippage or non-coincidence between touching and touched, affecting and affected, the perceiving body and the body as things among things" (Simonsen and Koefoed 2020, 29). In other words, the flesh, as the ontological realm of being within the fold from which the body-subject unfolds, precedes the subject and object as inherently reversible positionalities. But at the methodological-analytical level, the divergence between them, between the self and the world, remains, making the conceptual space for the body-subject.

Body-subject is born out of the "mass of the sensible" by *segregation*—it remains part of the sensible while being a different entity that appears as sentient corporeality (Merleau-Ponty 1968, 136). Merleau-Ponty articulates it as a relation of proximity and distance, which are not contrary to but consonant with each other (Merleau-Ponty 1968, 135). As Ted Toadvine (2009) states, the gap between the sensible and sentient aspects of the body—the coexistence of continuity and difference—is not a failure of Merleau-Ponty's thought; it is precisely in this gap that the world is disclosed. Indeed, it is this gap—*écart*, slippage, divergence—that provides a solid ground for exploring the ways in which we as human beings experience and relate to the world, both as a part of it and as different from it.

Consequently, this methodological difference Merleau-Ponty affords within his relational ontology makes it possible to account for the lived, embodied experiences of women, to explore "embodied being-in-the-world through modalities of sexual and gender difference" (Young 2005, 7), in relation to the material world that embodied subjects inhabit. In this respect, what Merleau-Ponty offers critical and feminist researchers who are interested in studying women's political subjectivity and agency is the conceptual vocabulary to embed the body-subject within the ontological realm of the flesh, without dissolving it altogether.

Yet Merleau-Ponty (1968, 248) maintains that the "flesh of my body is shared by the world," that the world "encroaches upon" my body and "it encroaches upon the world." To explore the ways in which we "encroach" upon the world, and the world upon us, however, the shared ontology of flesh can only be a starting point. We are of the world, but we are also in it; "folded" out of the same flesh, we still encounter the world and we respond to it with a certain operative intentionality (Rosa 2019; Oksala 2006) as embodied singularities. The body-subject is the embodiment of singularity, a site of lived, embodied perception and experience, of mobility and spatiality. It acts as an anchor, providing a certain point of view, an embedded and located perspective that is lacking in the new materialist and Deleuzian-inspired posthumanist ontologies (Fielding 2017; see also Kinkaid 2021). It is a reminder that sensations emerge out of more than human encounters, but are also "lived by particular people in particular places" (Clare 2019, xviii), a reminder that one should not lose sight of singularity within the posthuman universe of inter-/intrarelational, transcorporeal intensities, becomings, and visceral, vital forces.

EMOTIONS AND POLITICAL AGENCY

This chapter has explored bodily senses and affects as they emerge out of the bodily encounters of East Black Sea women, and some men, with river waters. One might add emotions to this picture, not as separate from, but in a continuum that also contains senses and affects[8] when it comes to connecting women's environmental activism to their embodied sensory relationship with river waters. Indeed, anti-HEPP activism seems to emerge within a continuum of bodily senses and affects, more articulated and coded emotions, and political agency. Here I understand emotion, following Brian Massumi (2002) and Deborah Gould (2009), as "captured" affect, as the "expression of affect in gesture and language" (Massumi quoted in Gould 2009, 20), as "one's personal expression of what one is feeling in a given movement, an expression that is structured by social convention, by culture" (Gould 2009, 20). Hence, "there is no emotion without bodily sensations, bodily resonance and affectability" (Fuchs 2013, 613).

In the case of the East Black Sea Region, as women's intimate sensory and affective connection with river waters is an essential part of their everyday lives as well as of their personal family histories, identities, and memories (as discussed in the next chapter), the prospect of losing the river triggers strong emotions such as anger, frustration, sadness, and despair (Norgaard and Reed 2017). This is why accounts of bodily connections, as presented above, are often followed by emotional responses and statements of determination. Let's take Semra, Gülşen, and Nuran, three women whose sensory connections with river waters are presented above, from two different valleys in Fındıklı, fighting to protect the Arılı and Çağlayan Rivers. Semra started our conversation with an emotion-laden proclamation, saying, "I am full [of emotions] for the river. I want to cry when I talk about it." She then told me, in detail, about her bodily connection with the river and her memories and family history as intertwined with the sight, sound, and touch of river waters (see above, and the next chapter). During our conversation, the initial emotion of sadness she expressed was transformed into frustration and anger when she started talking about the intransigence of HEPP companies despite the persistent opposition of the local community: "I have been keeping guard day and night by the river. I have gone to demonstrations. We have been making epic history here. And they are still after our river. How could they [dare]?" In the very next sentence, her anger gave way to commitment and determination: "We won't give up on our river. We won't! We won't! We won't!" [9]

Similarly, Gülşen and Nuran, who are both in their sixties, started our conversation by explaining how they felt when they heard about the HEPP project. Their initial feelings were shock and disbelief. Then came sadness and sorrow as they saw the damage the HEPPs had done in other places in their province: "Look at what they did to İkizdere [Valley]! It makes you want to cry." These emotions have not paralyzed them, however. On the contrary, they have derived from them the energy and strength to reinvent themselves as hard-core activists, fighting to protect the river they have listened to, looked at, swum in, and drunk from all their lives. It is the river that makes their village a special place, a place they always want to be ("I cannot sleep in the town center"), a place their children and grandchildren who live elsewhere dream of coming back to. The emotional compass then turned toward joy, excitement, and pride when they told me about their transformation from earthy housewives to well-known activists for the cause: "Now everybody knows us," they proudly said. As self-proclaimed "lawyers" and protectors of the river, they spoke in meetings, gave interviews to the press, walked at the front of demonstrations, sang songs they had written for the river, and attacked the vehicles owned by the HEPP companies with sticks and eggs. "I felt so strong when we were chasing them," one said. The other agreed: "We were actually quite gentle; we should have rolled their car over."

Emotions, as we know, are often associated with women, and this association is framed in a way to keep women (Åhäll 2018), as well as the rural and/or Indigenous populations (Velicu 2015), out of politics. This framing is employed every so often in environmental conflicts to discredit women's activism (Seager 1996; Bell 2013). We also know, however, thanks to the ever-growing literature of the last few decades, that the contrast between emotion and reason and/or intellect is misleading (see, e.g., De Sousa 1990; Hochschild 1998; Barbalet 2002), and that there is a close affinity between emotions and political action (see, e.g., Goodwin, Jasper, and Polletta 2009; Jasper 2011), which also applies to the field of place-based and/or environmental struggles (see, e.g., Woods et al. 2012; Raynes et al. 2016).[10] "In fact," as Celene Krauss, a scholar of women's environmental-community activism, states, "emotions may well be the quality that makes women so effective in these [environmental-community] movements" (Krauss 1998, 142). Michael Woods and his colleagues (2012) demonstrate how emotions drive and sustain political action in the empirical context of rural protests in the United Kingdom. Read together, the statements of the three women presented above illustrate what Woods and colleagues discussed as the "ladder of emotions."

Emotions of attachment to place, such as love, stand at the very base of their ladder. On this firm basis, emotions related to the perceived threat to that place, such as distress and anger, build up. These are then followed by the emotions reacting to the failure of the political system to represent the interests of the rural communities—such as frustration and alienation. This multiple and multilayered emotionality assists protest participation and strengthens the militancy of the struggle, which leads up the "ladder" to emotions such as excitement, thrill, and defiance. What the accounts of East Black Sea women also show us, with the abovementioned field of bodily sense and affect, emotion, and political action in mind, is the relational nature of sense, affect, and emotion and the ways in which they "bleed into" each other in conditioning embodied subjectivity and political agency. Despite the conceptual efforts to differentiate them clearly, emotions build on sensations, feelings, and bodily affects, and thus, "there is a sentient dimension to emotions" (Kruks 2001, 146; see also Åhäll 2018). Kay Milton (2002) explores this sentient dimension in her seminal book *Loving Nature*, showing how emotions emerge out of both social and ecological relations—out of our experience of living in the world. Loving nature, in this sense, is very much about having a direct experience of, and a sensory-affective connection to, nature and natural entities. And the emotions that grow out of this experience (i.e., loving nature) become the prime motivation for environmental activism and for caring about nature in general (Singh 2013).

In a similar vein, East Black Sea women draw very much on their love of rivers—emerging out of their sensory-affective connections with them— and their anger, sadness, and despair caused by the HEPP projects. "I didn't fall in love with my husband, but with the [Arılı] river," says Melihat Alişan, a woman in her seventies from Fındıklı (Kasapoğlu 2013). In another interview five years later, Melihat tells the striking story of getting up very early in the morning after her wedding night to go the river, leaving her husband behind in bed. "I love the river that much," she says. As the interview continues, we discover the roots of her love for the river, buried deep in bodily, sensory memories of her childhood: "My mother used to wash the laundry here in this river. She used to spread the sheets out on rocks to dry them. We children used to wait until they were dry. We used to go in the water to swim, wash, play.... We are children of the river." It is thus the intimate, sensory, and affective relationship Melihat describes that engenders the emotion of love toward the nonhuman, environmental entities and toward nature in the broader sense, much like love toward other human beings. And, as if to manifest the unbreakable connections between bodily senses and affects, emotions and political agency described above, Melihat ends the interview by emphasizing her determination to protect the river: "We are not giving our river waters to anyone. We put our lives on the line here" (Kalender and Başar 2018).

4 Place, Body, Memory

RIVER WATERS AND THE IMMANENCE OF
THE PAST IN THE PRESENT

Sultans, pashas decreed [the construction of the HEPP].
Will there be any peace if they lay their hands on the homeland [yurt] of the peasant? We didn't cut the trees. We
didn't burn the forests. We have founded this homeland
[yurt]. It is our village. It is our water.

Şerife Çalı, a seventy-five-year-old woman from the village
of Ahmetler (MR)

Şerife Çalı said this to a reporter from the daily newspaper *Cumhuriyet* (Yavuz 2014). The word she uses in Turkish, *yurt*, communicates a rich repertoire of meanings. In Western languages, the word stands for the round tents used by nomadic peoples of the Central Asian steppes. In Turkish, however, it has other meanings, such as "homeland" and "home country." Hence, it is a word that denotes a wide spectrum of deeply personal places, from the tent you sleep in to the country you were born into. One might say that the semantics of the word *yurt* signifies the relational nature of place, as it implies that place is not a singular unit of dwelling or a spatial infrastructure, nor a particular geographical location or a specific country, but a manifold relationship between different spaces, structures, localities, and scales.

Ahmetler is a Yörük settlement, like many other villages threatened by and resisting HEPP projects in the Mediterranean Region of Turkey. Yörüks were originally nomadic Turkmen communities that settled in the secluded but fertile valleys of the Mediterranean highlands. The myths and stories of settlement still survive among Yörük communities, as in the case of Ahmetler. Mustafa, a middle-aged man who is a retired teacher and a well-known anti-HEPP activist from Ahmetler village, told me about these stories and myths on a hot summer day in the town center of Antalya in 2014. As the myth goes, the village was established by three shepherds who shared the same name: Ahmet (*Ahmetler* is the plural of *Ahmet*). Each shepherd represents a different quality: dignity, justice, and courage (*asalet, adalet, cesaret,* respectively). They made use of those qualities as they revolted against the tyrannical Ottoman governor of the region, who seized the crops and cattle of the peasants by force. They then founded a new village in a secluded valley.

The founding myth of Ahmetler tells us a number of things. The very fact that such a myth has been transmitted down through the generations and survives as a constitutive element of identity and heritage among the villagers shows us how they relate to their village and valley as a place founded by their ancestors. It is thus a place that has been made by them for generations—it is their homeplace. Salih, a prominent anti-HEPP figure who was sixty at the time we spoke in the summer of 2014 in his village house in Karacaören, another Yörük village in the Mediterranean Region, expressed a similar claim: "We were born here. Our ancestors settled here after they migrated from Central Asia, who knows how many centuries ago. From that time till now, we have lived on this land with this river. We have inherited this place from our ancestors, from our great grandfathers." Having roots in a place is a key aspect that engenders attachment and belonging (see, e.g., Bell 2013). Another important aspect is the act of "making" the place—that is, the close, continuous, and habitual engagement with place that "enables a depth of attachment to that locality" (Strang 2005, 31).

Being embodied means, first and foremost, being situated—being *in place*. In that sense "being-in-place" is a key modalization of "being-in-the-world" (Casey 2000). Place, in other words, "is how the world presents itself" (Larsen and Johnson 2012a, 633). It is the immediate environment of the lived body, which is at once physical and historical, social and cultural (Whitmore 2019). Despite the obvious effects of global connections on the sense of place (Massey 1994), we are still oriented and situated in place as embodied human subjects: "Places are deeply historical and specific; no matter how transnationalized and shaped by larger forces, there is an important sense in which one always lives locally, in place" (Harcourt and Escobar 2005, 6).

What I aim to do in this chapter is to expand the body-centered, phenomenological framework I have been developing in the previous chapters by establishing a relationship between water, place, body, and memory.[1] I demonstrate throughout the chapter how people experience this relationship and how these experiences drive their struggles against HEPPs. Establishing this relationship between river waters, place, body, and memory, I draw on phenomenological and Indigenous conceptions of place as a field of more-than-human relations and "as a way of knowing, experiencing and relating with the world" (Coulthard 2010, 79). Developing the discussion, I introduce the concepts of body memory and place memory with reference to the works of Edward Casey, Thomas Fuchs, and Maurice Merleau-Ponty. The chapter also aims to demonstrate the conceptual potential of a body-centered approach to studying the relationship between water, place, and identity, which is central to grassroots struggles for water bodies such as rivers (see, e.g., Espeland 1998; Sylvain 2002; Boelens, Gil, and Getches 2010; Gelles 2010; Boelens 2014). Instead of establishing this relationship through culture and/or cosmology, in which water bodies as sacred entities are constitutive of cultural identities, I put the lived body at the center of the analysis, building on extensive empirical research and a phenomenological discussion of body, place, and memory.

RIVER, PLACE, AND IDENTITY: ANTI-HEPP STRUGGLE AND THE MORE-THAN-HUMAN MAKING OF PLACES

Places are just as constitutive of our sense of self and identity as our experiences and practices are constitutive of places. This mutually constitutive relationship is most visible in local environmental struggles. Local communities make the villages and valleys where they live into places by means of habitation and dwelling, and through various social and economic practices that sustain habitation and dwelling. Rural and Indigenous communities in particular live in places of their own making, often for generations, and they claim these places as their own. Their experience of place, hence, is grounded, foundational, and "thick" (Hay 1998), in contrast to the supposedly "thinned out" (Sack 1997) sense of place of the modern, cosmopolitan subject as an ideal type who lives across many places in a global world.[2] This foundational and "thick" sense of place provides a solid ground on which to claim the place as one's own, and fuels struggles to protect this claim against the claims of the state and/or private companies.

Returning to the Ahmetler example, another aspect of its founding myth is the narrative of the tyrant state (or its representatives) and the resisting peasants, which is being repeated in the case of the imposed HEPP projects and rural resistance against them. Ahmetler witnessed a violent conflict between the villagers and the HEPP companies—three companies in all, as one company withdrew and another took the project over. During this long process, villagers physically blocked the construction work and ultimately deterred the first two companies. The third company, however, used physical violence against the villagers and injured three of them with real bullets. Gendarmerie intervened in the next physical conflict between villagers and company workers and representatives, a week after the first incident, only to attack the villagers, battering men and women. In the end, the villagers won the struggle, as the project was suspended by a court decision and the third company also withdrew. In Ottoman times governors seized whatever little property and products peasants had in the form of taxes; now it is private companies that come and seize the flowing waters, the rights to which they "bought" from the state. Very much like the Ottoman governors, the companies are protected by soldiers and use physical force against the peasants. And once again, peasants find themselves in the position of rebels, bandits, criminals.

What is interesting here is that the same experience of violent dispossession and rebellion is repeated yet again, despite the fact that they settled in the remote, secluded, mountainous parts of Anatolia. They were used to being out of sight and out of mind vis-à-vis the state, until their villages were targeted by HEPP projects. The steep valleys where their villages are located, with their rivers and streams flowing along elevated streambeds, proved to be the perfect locations for such projects. Once again, the state was coming to take, despite having failed to give. Refiye and Hamza, an elderly couple from the village of Yaylacılar (EBR), expressed this very feeling when I talked to them in front of their house in the summer of 2013:

> This is our village. This is where we live, where we settled. We don't want anything from the state. They did not even build us a road to the plateau [*yayla*]. We walk eight hours to get there. Now they want to take our water. They cannot. We do not want their road; we do not want their money. And we will not give our stream.

It was also hard to believe that those home-places, which are imagined to be stable shelters that resist the rapid spatial transformations that urbanized areas of the country have been undergoing,[3] were becoming targets of similar processes. Halime is a middle-aged woman who became a symbol of what came to be known as the "yellow scarf" (*sarı yazma*) movement in Loç Valley (WBR). When she was younger, she and her husband moved to Sultanbeyli, a working-class suburb on the outskirts of Istanbul, to earn a living. It was always clear to her that she would return to Loç, which is located by the Devrekani stream. She told Hilal Köse, a reporter from the daily newspaper *Cumhuriyet*, how she had raised her children by that stream (Köse 2016). Her boss in Istanbul saw a picture of her children swimming in the stream (probably taken on one of their summer trips back to the village). To her surprise, her boss said all this landscape would be ruined soon; the fresh waters would be seized and appropriated. At the time, she did not take his comment seriously, as she believed that nobody could even find the remote and secluded Loç Valley, their "secret garden," hidden away among the Küre Mountains of the West Black Sea Region.

They did find it one day, though. As no one gave the villagers notice or informed them of the project, it felt like an invasion when the company representatives and workers suddenly appeared with construction equipment in the valley. When the locals started to organize against the project, Halime became a leading activist, writing to state institutions, filing lawsuits, blocking the construction site, and leading demonstrations. She even spoke in the European Parliament in 2010 to present the claims of her community to an international audience that included Nicholas Hanley, head of the European Commission's International Relations and Enlargement Unit and of the Directorate-General for Environment at the time.[4] She stated that protecting her place, of which the stream is an

integral part, was the main motivation behind her struggle, which ultimately led to her personal bankruptcy and to legal charges being brought
against her. She says she remembered what her grandfather had told her
when she was a child as she faced the prospect of the destruction of her
home-place: that she had a right to this place, to its mountains, rocks, and
streams (Köse 2016).

Halime's case demonstrates how her claim on her homeplace has driven
her political agency against HEPPs. Her story is also illustrative of the
shock and distress that the HEPP projects induce, and the hardships
resisting communities endure. Halime had to stand trial for "slowing
down the construction of the Cide hydropower plant," along with eighty-
three other residents of Loç Valley, which included women over eighty
years old. The incident was ironic, as the Kastamonu Administrative
Court had already stopped work on the project once. The construction
started again after the Council of State (Danıştay), the highest administrative court in Turkey, overturned the decision. When the residents were
taken into custody and faced legal charges, the project was at the administrative court for the second round of trials. In the end, the court once
again suspended work on the project, and this time the decision was rati-
fied by the Council of State.

It is well established, especially in the field of environmental psychology, that people develop "place attachment"—strong emotional bonds
with places (see, e.g., Altman and Low 1992; Hay 1998; Hidalgo and
Hernández 2001; Manzo and Devine-Wright 2014) and that they have
"place identity," meaning that they deeply identify with certain places (see,
e.g., Proshansky 1978; Stedman 2002; Waster-Herber 2004). Place
attachment and place identity are strong drivers of place-based struggles,
especially in Indigenous and rural settings.[5] As Halime's story shows,
place attachment and place identity are not limited to residents who never
left a given place. The remote, mountainous, rural areas, which are seen as
perfect settings for HEPP projects, are also places of emigration. The
Black Sea and the East and Southeast Anatolia Regions are known as the
principal regions of origin for migrants fueling the process of migration
from rural areas to urban centers that started in the 1950s and continues
to this day. Migration, however, cannot dissolve the strong sense of connection with the home-place, and sometimes even strengthens it.

Murat Öztürk, Joost Jongerden, and Andy Hilton (2021) stress the centrality of homeland to social identity in Turkey. "Even those who are permanently resident in the city," they claim, "psychologically reference and routinely cite their community of origin" (Öztürk, Jongerden, and Hilton 2021, 331). They demonstrate how villagers who migrate to urban areas maintain "the rural social fabric but from a distance" (Öztürk, Jongerden, and Hilton 2021, 330). Only one in seventeen people who move to the city sell any of their lands in the village (Öztürk, Jongerden, and Hilton 2018). Maintenance of the rural social fabric involves keeping the land and returning to the village for holidays, family visits, important events such as weddings, and, of course, to help with the harvest. Development of mobility and communication technologies is increasingly making possible hybrid living structures that include maintaining "multiple places in the village, in the local town and the nearest city, in the distant metropolis(es) and foreign countries" (Öztürk, Jongerden, and Hilton 2021, 331). It is indeed common among migrants from the Black Sea Region to keep a house in their home villages so they can spend their summer holidays there and engage in the tea and hazelnut harvest—"the phenomenon of the summer village" (Öztürk, Jongerden, and Hilton 2018, 252). Many dream of retiring to their homelands, and they make sure that their children have a connection with the place, carefully cultivating a strong idea of roots and a sense of belonging to the place as their *yurt*. As Selime, whom I cited in the introduction, told me as we sat on her terrace in the summer of 2013, "I have been in Istanbul for forty years. But I built a house here in my village. Our parents left this place to us, and we want to hand it down to our children, as it is. We would not exchange this place for anyplace else. Even if they gave us Etiler, Sarıyer, and Caddebostan [upscale suburbs of Istanbul], we would still come to this very village and live here."

This strong connection to the homeland underpinned by a dual-life structure—which is depicted by terms such as *rural–urban connectedness* (Öztürk, Jongerden, and Hilton 2021) and *urban–rural continuum* (Erensü in Evren 2022)—is key to understanding why protecting the home-place and the river is an important motivation for people who live in the city. Their connection and connectedness are related to the widespread understanding of the homeland—the village—as a shelter, as the incommensurable "original place" from which they come and to which they will return.[6] Sevcan, a young woman from Aslandere (EBR), made this point while she and her sister Asiye and I were having a cup of tea: "We have a splendid place here. We don't want to lose it. Wherever we go, we want to come back and live here in our village." Öztürk, Jongerden, and Hilton (2018, 251) make an important point about the strong connection people have with their homeland in Turkey. They maintain that the home and land cannot be separated from their "ecological setting in the local human and natural environment," which is central to the lived experience of people who live (partly or fully) in the villages. Rivers are the main features of this "ecological setting" in many parts of Anatolia, particularly in the East Black Sea Region. The central importance of rivers to (home-) places is manifested in the strong belief among local people that the place will cease to be *their* place—it will be spoiled, it will not be the same, it will not be livable—if and when the river no longer flows through it.

Even though the specific arguments and motivations for opposing hydropower might differ from one place to another, statements such as "we cannot live here without the rivers" are commonly heard in many local communities, regardless of their differences. In some places, the meaning is quite literal, as river waters sustain livelihoods. In other places, however, these statements imply more than immediate dependence. Rather, they indicate the central importance of the rivers for the maintenance of the *sense of place*, that is, the felt qualities of a given place that make it distinctive (Malpas 2018; see also Buttimer 1980). This is what Asiye indicates when she says: "We don't want them to touch our rivers. From our ancestors to our children, we have been living here by this river [the Çağlayan]. It is where we live, it is our life. No one has the right or the power to ruin my place, my nest, my homeland." Asiye, like many others in the region, equates the loss of river waters with the loss of her home-place. Thus, the loss can be experienced not just when people are displaced because the place has ceased to exist, as in the case of hydro dams, but also when the place is stripped of the qualities that make it distinctive and, as Heidegger (1996) would say, *heimisch* ("homey").

Rivers make places and shape the sense of place; they configure not only the geography and the landscape, but also the built environment, as their flow configures human settlements (see, e.g., Kelly et al. 2017; Stefanovic 2020b) and characterizes the habitual behaviors and everyday practices of those who settle around them.[7] The natural landscape of the East Black Sea Region, for instance, which has been devastated in the seaside town centers due to haphazard urbanization, unfolds to breathtaking sights in the villages and valleys that hide away at the foot of the Kaçkar Mountains, and, in its most imposing form, in the plateaus and summer pastures (*yayla*) that are located high up in the mountains. These villages, valleys, and plateaus never fail to impress with their multishaded greenery covering the hills surrounding the settlements. Fields of tea plants contribute to the many shades of green that characterize the landscape. The rivers are the crown jewels in this landscape, bringing life, texture, and character to the densely woven vegetation. It seems that what makes those villages "a heaven on Earth," as many villagers say, is above all the cascading flow of the streams. This is the central element of the villages' aesthetic and affective atmosphere; it goes hand in hand with the rivers' refreshing, cooling, and reviving qualities, and is fundamental to the production of the widespread image of Black Sea villages as splendid places, as little slices of heaven on Earth.

This representation of the typical East Black Sea village is reminiscent of descriptions of heaven (*Jannah* in Arabic; *cennet* in Turkish) in the Qur'an. The word *jannah* itself means "garden" in Arabic. Consequently, heaven is represented as a garden through which rivers flow: "But give glad tidings to those who believe and work righteousness, that their portion is gardens, beneath which rivers flow" (Qur'an 2:25). "Allah promised to believers, men and women, gardens under which rivers flow, to dwell therein, and beautiful mansions in gardens of everlasting bliss" (Qur'an 9:72). Invoking these descriptions in the Qur'an, Hüseyin Paker, the imam of the village of Camili (EBR) in the geographical area of Macahel (Machakheli), which Turkey shares with Georgia, says, "If you are looking for heaven (*cennet*) on earth, Macahel is it. But they want to turn it into a hell with HEPP projects" (Hamsici 2010, 201–2).

This centrality of the rivers' flow to the sense of place became clear during my fieldwork in the region, as I heard, over and over again, how people "escape" to the villages in summer. For families who have a second (or first) residence in large cities or in regional town centers, villages act as a refuge from the noise, heat, and general hustle and bustle of the urban areas. Many say that they cannot stay "down" (regional town centers are often at sea level) because of the heat and humidity. "I was down in Fındıklı [town center] yesterday but could not stay overnight. I ran back to the village," Nuran, an elderly woman from Aslandere, told me. Refiye expressed a similar experience: "I was down in the town center. I could not stay for even two hours. I came back to the village. Excellent!" Refiye also stressed that the cool, fresh air of the village is not only due to its altitude, but also to the Arılı River running through the village. While we were sitting on her doorstep, she said, "We cannot sit here comfortably if this river does not flow. If the river is gone, then this cool, fresh air is gone too. If they take our stream from us, they may as well just put us into our graves alive."

There is surely a psychological aspect to this intimate relationship between place and identity, as the concepts of place attachment and place identity attest. We tend to think of ourselves as defined in reference to particular places, mostly to the places where we were born and raised or where we spent a formative time of our lives. Intertwinement of place and identity, though, is not only an issue of self-identification, as Malpas (2018) reminds us. Our identities are place-bound, as they are formed within our active, daily engagement with the surrounding environment and our embodied encounters with human and nonhuman others. Place, in this sense, is made and remade through "encounter, dialogue and relationship among humans and non-humans who share the landscape" (Larsen and Johnson 2017, 2); it is "the relations of human and other-than-human beings that interact in a given territory marking it as a specific place" (de la Cadena 2010, 353). Indigenous conceptions of place as encompassing all of the "people and animals, rocks and trees, lakes and rivers, and so on" that share a landscape or a territory (Coulthard 2010, 80; see also de la Cadena 2010; Larsen and Johnson 2017) are particularly helpful here to conceptualize place not just as a location in which we live, but as a world of bodily connections and socio-material relations that shape us while simultaneously being shaped by us.

From such a perspective, it becomes possible to perceive the rivers not as a backdrop to the place, but as constitutive of it. In the case of rivers in the East Black Sea Region, it is not only the spatial agency of the rivers in configuring human settlements that makes the place; the rivers' place-making agency also involves their role in shaping the place-world of bodily connections and social-material relations. Take Meryem Demircan, a middle-aged woman from the village of Rüzgarlı in İkizdere Valley, who told a reporter from the daily newspaper *Milliyet:*

> We won't give up on this [river] water. Why? We were born here, grew up here. . . . This water is our love, our parents, and our source of life. . . . Now this village is 400 years old. How can you take our water of 400 years just like that? (Ünal 2013)

Watching the documentary film *Bir Avuç Cesur İnsan* ("A Few Brave People"), we get a glimpse of Meryem's life in her village and in the Kotaflar plateau (*yayla*) where she spends her summers with her goats. We see the breathtaking views of her house on the plateau, which stands above the clouds. We also observe her compassionate relationship with her goats—the way she talks to them as companions, and the way she shepherds them through the natural landscape of the plateau. Later in the documentary, we also see her talking about her opposition to HEPPs in her simple house in the village of Rüzgarlı while holding a kid in her lap. The camera follows her to the Rüzgarlı river as she jumps from one rock to the other in the river, as she washes her face in the fast-flowing, crystal clear waters of the river, and as she drinks from it. The whole sequence makes it clear that the river is an essential element of her sense of place due to its primacy in the place-world she shares with her goats, which comprises the village and the plateau and the river that links the two.

Micropractices of everyday life connect the geographical subject to her place-world (Casey 2001). In the case of Meryem, it is the daily bodily, sensory, and affective encounters—seeing, hearing, touching, tasting the river—and the affective registers these sensory encounters engender that accompany her routine and habitual practices, that ultimately create the unassailable association of the river with her home-place. Meryem not only talks about the history of the place in the comment quoted above; she also refers to her personal history and memory, and associates the river with love and with her parents, indicating the intimacy of her relationship with it. Meryem's references to her personal history and memory become particularly relevant when read through this corporeal relationship between river waters and personal histories, memories, and heritage, as I detail in the next section. Her associations of the river with love (and lovers) and her parents, on the other hand, will be further developed in chapter 5.

MEMORIES OF THE BODY, MEMORIES OF THE PLACE: RETRIEVING THE PAST THROUGH EMBODIED SENSORY INTERACTIONS WITH RIVER WATERS

Everyday, even banal personal memories of persons, events, and places remain underrepresented within the field of memory studies, which is mainly occupied with collective and social memories (Jones and Garde-Hansen 2012a). Memories of everyday life and experiences, however, are central to our sense of self and our attachment to place. When we talk about memory, we refer to a mental capacity—to remember and reclaim, to retain and retrieve. We mostly overlook the ways in which this act of recollection occurs within and through the corporeal-material interaction of our bodies with our environments. Bodies are not only central to how we experience the world; they are also fundamental to how we store and reclaim those experiences as memories. Memory, in this sense, can be defined as "a process of encoding and storing records of experience which can be retrieved or which re-emerge in subsequent practice" (Jones and Garde-Hansen 2012b, 20), which involves lived bodies, things, and environments situated in place.

The concept of body memory, on the other hand, is largely associated with implicit memory, what Merleau-Ponty defined as "knowledge in the hands."[8] Body memory, however, is not limited to this habitual type, which is widely discussed through well-known examples of riding a bike or playing piano as implicit recollection of pre-reflexive knowledge that is embodied in and expressed through physical activities. Body memory entails a broader range of memories of "being bodily in the world" as "instances of remembering places, events and people with and in the lived body" (Casey 2000, xi). Casey is in line with Merleau-Ponty's concept of the body-subject when he defines body memory as a "trace," as "a survival of the past, an enjambment" in his published course notes (2003, 276). The past, in other words, exists in the present in bodies, through body memory (Casey 2000). In the case of the East Black Sea Region, corporeal survival of the past, in the bodies of people and in the waters of the river,[9] is constitutive of their political subjectivities as they protest against hydropower plant development.

Thomas Fuchs (2012, 11) maintains that corporeal experiences anchored in body memory "spread out and connect with the environment like an invisible network, which relates us to things and to people." Body memories extend to spaces, places, and situations (Casey 2000 and 2001; Fuchs 2012). They are very much entangled with the tangible materiality of our place-world (Jones and Garde-Hansen 2012b). Memory is, then, like perception, not the act of remembering that occurs in an isolated mind; it is a complex intercorporeal and situational process whereby places, bodies, and things reenact the past. It is not limited to a set of abstract recollections insofar as the past is not *represented*, but is *reenacted*, as Bergson once put it, through body memory, through the everyday sensory relationship with river waters. Thus, we access the past not so much through images and words, but primarily through immediate experience and action, in place.

The example both Thomas Fuchs (2012) and Hartmut Rosa (2019) invoke is the famous madeleine episode in Marcel Proust's *In Search of Lost Time*. The episode, as is well known, illustrates how a piece of tea-soaked cake brings complex memories back. It demonstrates how familiar sensations aroused by certain spaces, places, situations, and bodily encounters can function as *memory cores* that recall and release enclosed memories under suitable circumstances. Fuchs (2012) discusses this in terms of "situational" and "intercorporeal" memory. I encountered various examples of situational and intercorporeal memory when I asked villagers about their motivations behind opposing hydropower projects. Particularly in the case of middle-aged and older residents, the memories of their childhood and their parents often came up. Those memories often involved their bodily-sensory connection with river waters. Semra explained to me how her memories and personal history are inextricably bound up with the river waters in front of her house: "I see my mother and my father by this [Arılı] river; every time I look at the river I remember them. . . . We are keen on our history. How could I give up on *my* river?" The first sentence of Semra's statement clarifies how memory cores work. The visual memory of her mother and her father by the river is released every time she sees the river in the present.

Sight, of course, is not the only sense that possesses the capacity to revive the past. All the familiar sensations of bodily experiences that involve river waters—the sound, the taste, and the touch of the river—also function as memory cores, reenacting past events and situations in the present. Orhan Kemal Topal, now a middle-aged man, tells the producers of the documentary film *Akıntıya Karşı* ("Against the Current")[10] how he used to take the family's cows to the *yayla* when he was about twelve years old, and how he used to drink water from the very source of the Çağlayan stream. Drinking from the source of the stream is a practice he repeated countless times in his life, through which he reenacted the memory of his early youth. Touch, which appears to be a powerful body memory, can also function as a memory core, as manifested in the narratives of playing, washing, and swimming in the river waters. Ayşe Pervaz, a young woman at the time of the interview, tells a reporter from the daily newspaper *Evrensel:* "We spent our childhood here in the river. We learned how to swim in this river. We used to harvest tea with our families, but around three o'clock in the afternoon all the young people of the village [Konaklı—EBR] used to gather by the river. Then we would swim together (Uğurlu 2015)."[11] In the case of sound, memories are often related to hearing the river during childhood and growing up with that sound. Süleyman Bilgi is a young man from İkizdere Valley (EBR), which is being destroyed by multiple HEPP projects. He expresses this common theme in the documentary film *Bir Avuç Cesur İnsan* ("A Few Brave People"): "The sound of the [İkizdere] river is like my mother's lullaby; I grew up with it. It is a sound we have heard since our childhood. It is always in my ears. Hearing it gives a sense of peace. How could we live here without it?" The way he compares the sound of the stream to his mother's lullaby not only refers to the calming, relaxing impact of it; it also implies that the sound of flowing water is entwined with the most intimate sounds of his childhood. Süleyman also says, in the same documentary, "Without this river, I cannot remain who I am. I cannot continue living in this village," revealing the close affinity between sensory body memories of childhood, the nonhuman entities in which those memories are anchored, and sense of self and identity. His words proclaim that memories of bodily experiences act as tissues of identity (see Jones and Garde-Hansen 2012a). Identity is understood here not only as a fixed category assigned to group belonging, but as a dynamic process of becoming performatively constructed through a range of embodied experiences, and memories of them, as "conglomerations of past everyday experiences, including their spatial textures and affective registers" (Jones and Garde-Hansen 2012a, 8). Personal identity, in this sense, is "rooted ultimately in body memory" (Casey 2000, 176). Thomas Fuchs (2012, 20) articulates this in a compelling way: "[B]ody memory is the underlying carrier of our life history, and eventually of our whole being-in-the-world."

"Senses are powerful sources of body memory," Susan Steward (2005, 59) tells us, due to the body's capacity to carry memories of senses somatically. Oftentimes our sensory memories, which are embodied registers of our encounters with the sensible world, involve more than one sense in complex interaction. As they are anchored in place, memories of senses are situational memories, to use Fuchs's typology. And, as in the example of the madeleine, different senses participate in various combinations in situational memory. Situational memory emerges through the interaction of bodies, places, and things under certain circumstances (Fuchs 2012), and reflects the multisensory character of those everyday interactions. Let us take the example of Selime, whom I mentioned above. Selime, who was around sixty years old at the time we talked, told me that she was ready to die to protect the river, and pointed to a spot several hundred meters ahead along the riverbed: "Look, there is a waterfall over there. We used to take a break here with my parents when we were on our way to the plateau [yayla]. We would always eat and sleep next to the river during our journey." Every time Selime sees that waterfall, or walks again along the river to the plateau, every time she sleeps and wakes up—as her house is located by the river—to the sound of it, she remembers her childhood, her parents, and the intimacy she shared with them, which are anchored in place and, most of all, in the very flow of the river.

Memories of the past localized in the body become the present (Olkowski 2017). For Selime, it is the multisensory body memories, as well as her embodied experiences of river waters in the present, that retrieve those memories, that connect her, now middle-aged, to rivers and to her long-gone childhood and ancestors. Ayten Aydın, from Arhavi (EBR), also talks about her childhood memories when she is asked about her opposition to HEPPs. Her memories are also multisensory body memories of river waters:

> Our mothers worked in the fields. We, the children, used to go down to the
> river and spend all day there. The river was our playground. We used to
> drink from it, eat our meals by it, and wash ourselves with its waters. We will
> fight to the end, and we won't give up on our river. (Uğurlu 2015)

Bodies bear the traces of the places in which they are situated in many different ways. Those traces provide a ground for recollection, even when the
specific bodily encounters through which body memory is produced, and
which are contingent upon a specific combination of temporal and spatial
aspects of place, are no longer available. As Casey (2001, 688) puts it,
"The presence of place remains lodged in our body long after we left it;
this presence is held within the body in a virtual state, ready to be revived
when the appropriate impression or sensation arises." This idea of the
body as a trace, of the past and of the place, an idea originally belonging to
Merleau-Ponty and developed by Casey, is manifested in the narratives of
the villagers who must live away from their villages, and thus their rivers,
for reasons of work and family. Semra's statements are exemplary in this
regard. Semra, whom I mentioned above, is a retired nurse who, for family
reasons, divides her time between Istanbul and her village (Gürsu). She
describes the experience of being away from the village, from the river: "I
lie down on my couch in my flat in Istanbul, but my feet are always in the
waters of my river. Seriously. I'm speaking from my heart. The river is
always with me, in my mind, in my imagination, in my dreams." This narrative of prolonged bodily connection, of feeling her feet in water, hence
the reawakening of the sense of touch based on body memory, reminds us
of Merleau-Ponty's much-quoted passage on phantom limbs (2012,
78–89). Casey (2000) discusses the phenomenon in terms of the customary body's dissociation from the momentary body.[12] In one case, one still
feels a limb that is not there; in the other, a person (Semra) feels her feet
in the water, which is not there. Semra's case, then, could be seen as an
extended case of the phantom limb, where the sense of body is extended
to the waters of the river through the sensory connection between human
bodies and bodies of water. Other women from other parts of the world
also talk about such prolonged sensory connections with places and/or
nonhuman entities that hold even in the absence of those places and entities. Shanti, who lives in the North Karanpura Valley in eastern India and

whose story is told by Lahiri-Dutt (2015, 50), articulates a similar sentiment:

> [The spring] was a place where we gathered every day to bathe our children and ourselves. . . . One morning TISCO company bulldozed that spring, and with it the large *asan* tree next to it. . . . I can still see the spring and the *asan* tree next to it, if I close my eyes.

Place is selective for memories and memories are selective for place (Casey 2000). In other words, memories "seek out particular places as their natural habitats," and places guard, keep, and "hold in" memories (Casey 2000, 189). In the examples above, the "natural habitat" of the villagers' memories—of their childhood and youth, of their ancestors and other loved ones, of the important events of their lives—is the river. The river guards the memories and life histories; it holds them in place. Malpas (2018) discusses, with reference to Faulkner and Proust, how the temporal is spatialized and materialized—"memorized"—in the objects around us in the present. The river, in this sense, is the materialization of the past. The past leaks into the present with the flow of the river, as childhood experiences are retrieved within the constant everyday interaction and sensory connection with its waters. The loss of the river might not mean the loss of memories per se, as mental constructs. But it means the loss of the "intimate relation between memory and place, realized through the lived body," and the loss of the "immanence of the past in the present" (Casey 2000, 168). What Casey means by this expression is how body memory allows "the past to enter into the very present in which our remembering is taking place" (Casey 2000).

The capacity of the lived body's reenactment of the past (Bergson [1896] 1988), a process Casey defines as re-entering, in-sertion, and import of the past to the present, is very much anchored in place. When the connection between body, memory, and place is broken, the capacity of body memory to keep the past in the present is lost. It is only "moving in or through a given place" that "the body imports its own implaced past into its present experience" (Casey 2000, 194). Place, in this sense, acts as the condition of the continuance of time, of the presence of the past. Thus, struggles to protect a place should also be seen as struggles to protect the very condition of reenacting the past, to keep it alive in the present in place. Bell (2013, 124) cites Lorelei Scarboro, a Central Appalachian activist against mountaintop mining, who says:

> It's difficult to explain the attachment, the sense of place that Appalachians have. It's a connectedness to the land, to your surroundings. It's not the value of the house, it's not the price of the ten acres. It's the memories. It's what you have there. It's the life you share with the people you love.

Patti, a Native Yavapai woman who fought against construction of the Orme Dam in Arizona, picks up where Lorelei leaves off:

> [The struggle] is about this land. . . . A lot of things have happened here. The people lived here. They lived off the land. They died here. We all remember that so and so used to plant here, or sit over here, walk this way. We go through this all the time. I do that with some of the elderly. They tell what it was like, where people lived. . . . We remember with the land. (Espeland 1998, 200)

This act of remembering with the land through bodily actions and interactions is not limited to happy memories that are bound up in place; it also involves memories of hardships and sad events. Cemal Can, one of the opponents of the (thus far blocked) HEPP project in the Loç Valley (WBR), says in the documentary film *Sudaki Suretler* ("Figures in the Water") that they used to grow linen when he was young, and explains, at length, the hardships of weaving linen to make a pair of underwear. The process of separating the parts of the plant that are used for making thread from the rest of the plant involved soaking the plant in the waters of the river for about a month. This is why the river constantly reminds him of the hardships he endured in the past.[13] The memories of these hardships, though, are not disturbing memories to be repressed; they are valuable instances of life to be cherished and protected. Thus, he looks at the river *because* of the hardships he endured, because they remind him of memories he wants to hold onto. When I look at the same river (the Devrekani), I also see how beautiful it is. I see how the instantly striking and impressive Loç Valley would be empty and naked without its stream. But Cemal sees something beyond its beauty; he sees his past in the water. As Raymond Williams articulates, "The landscape takes on a different quality if you are one of those who remember. The scenery . . . is never separate from the history of the place, from the feeling for the lives that have been lived there (Williams, quoted in Jones and Garde-Hansen 2012c, 87). Memory intertwines temporality and spatiality, as we remember the past in place. In the example of rivers, one might understand why it becomes the symbol of the place, as the primary nonhuman entity that holds the past in place. If Edward S. Casey (2001), along with many other philosophers and geographers, is right about the co-constituting feature of the (geographical) self and (lived) place, it becomes clear that such a radical alteration of place—think about the sudden disappearance of a stream that has always flowed through the village from the mountains to the sea—triggers a disruption in the sense of the self as well. Places, even rural and relatively secluded places, change at a fast pace in modern times. Black Sea villages are also changing, as new houses (in most cases bigger and less aesthetically pleasing than the old ones) are built, new fields are enclosed for tea plantations, and guest houses and bungalows for tourists pop up as tourism slowly infiltrates the regions. Within this changing landscape, the river is the one constant, the stable aspect that stands for the continuity of the past and of life itself. Lewicka (2008) stresses that place identity encompasses aspects of continuity (sameness) and distinctiveness (uniqueness), which are the two literal meanings of the word *identity*. Rivers materialize both of these aspects, standing for continuity with the past and the distinctiveness of the place.

CONTINUITY AND FLOW: RIVER AS FAMILY HERITAGE

> I was born by the river. My mother washed me in that
> beautiful river for the first time.
> Melahat Alişan, a woman in her seventies from Çağlayan Valley (EBR),
> quoted in Kasapoğlu 2013

> We grew up with this river our whole lives. My grandmother,
> her father, our ancestors . . . all grew up with this river. How
> could I lose my river? How could I destroy all the memories
> left to us?
> Selime, a middle-aged woman from Arılı (EBR)

The theme of continuity becomes especially clear in the narratives of older women. Elderly women frequently mention their lifelong connection with river waters, saying, "We were born by this river and we will die by it, with its sound." Another common saying is "Rivers will flow until eternity," which implies that any rupture in the flow will also interrupt the continuity of life itself. By sustaining the continuity of life, the river's flow connects past and future generations. As seen above, memories of childhood, of parents and ancestors, are very much bound up with the river, as the defining element of the place. Associating the river waters with parents and ancestors is very common; persons and places are often bound up in memory, as one often recalls "not only the person, but the person and place, and both as part of the same image, part of a single remembrance" (Malpas 2018, 179). Memories of the autobiographical component in particular, Malpas (2018) stresses, are typically anchored in particular places, and the memories of them are foundational for our sense of self and identity. Consequently, the commitment to protect the rivers seems to be closely connected with villagers', especially women's, determination to keep the place, and the childhood and family memories bound up with that place, intact.

Keeping "the past in place" intact is not the only motivation for protecting the rivers. In line with the theme of continuity, it is also transferring the river to the next generation. River waters manifest the continuity of past, present, and future, intertwining them in place. For many villagers, it is not only a house or a small piece of land inherited from the ancestors; the place—the village, the valley—and the river as the defining feature of the place are a part of the heritage that should be left to the coming generations. It is thus not surprising to observe a common theme of children and grandchildren, the coming generations, as those for whom the river should be protected, preserved, and to whom it should be transferred. Not exclusively, but especially women, mothers, and grandmothers stress this aspect repeatedly. For these women, it is clear that rivers are loaded with memories. But they are also the carriers of those memories, of life histories, and of the cultural heritage for the coming generations. Ayşegül is a young woman with a generous smile from the village of Konaklı (EBR). She speaks fondly of her memories of the river: "We grew up with the river, swam in the river, we learned how to fish. Now our children are

Figure 3. Children playing in the Arılı River. Photo by the author.

learning how to swim and fish." What Ayşegül indicates is that rivers flow from the past to the future, connecting the bodily experiences and memories of the ancestors to the experiences of the children. Women mention their children's corporeal experiences with river waters (swimming, fishing, playing in water) as an important moment of heritage transfer.

Rivers also serve as playgrounds in those remote villages where no playground exists. I observed many children playing in and by the rivers in the East Black Sea Region and listened to many women who assured me of their children's and grandchildren's fondness of rivers. Elif told me during our conversation by the Arılı River that her child had just learned to swim in one of the pools that forms in the bends of the river. "He was very excited that he could swim in the deep parts. If it was not for this river, where could he swim?"[14] Playing and swimming in river waters in the summer is a much-beloved activity, not only for the children who permanently live in the villages, but also (and perhaps especially) for those who live most of the year in Turkey's big cities and in the regional towns along the coast and who have very limited experience of nature. Mahmut Hamsici cites Güllü Gül, a young woman who says, "The holidays started [in Istanbul] on that Friday. We arrived in the village [in the Loç Valley (WBR)] that very same evening. I have an eight-year-old daughter. She was just dying to go to the river. You know when you promise to buy a bicycle for your child if she studies well? Well, we promise her the river" (Hamsici 2010, 31). Similarly, Refiye, whom I mentioned before, pointed out three young children in the [Arılı] river, shouting and playing with joy during our interview. "They are my grandchildren," she said. "Look how happy they are in the river." Nuran, who was in her late sixties when I spoke with her in her village, Aslandere (EBR), told a similar story of her grandson, who often calls from Istanbul to tell her that he misses the river. "Grandma, I will come very soon," he tells her. "We will go down the river with my dad to fish."

Such narratives make it clear that rivers play an important role in social reproduction in the sense of providing a free, natural space for children to swim, fish, and play together—in short, to grow up together. Yet, as noted above, women's motivation to protect the river as a space for children goes beyond the rivers' role as playgrounds. They also strive to transfer a certain cultural and family heritage to their children in the form of corporeal experience and sensory-affective connection. As in the case of memories, the loss of river waters might not dissolve that family heritage per se. But it would break an important aspect of corporeal and experiential continuity, as the children would not be able to share the same bodily experience of river waters. Duygu Kaya, a young woman from Borçka (EBR), tells the story of such a loss:

I live in the house across the stream. I grew up in that house. Our stream was so beautiful. We used to play in it all the time. You could swim, you could fish. Now [after the HEPP project was constructed] it hurts me to look at it. There is hardly any water. It has become like a garbage dump. It stinks and it's a breeding ground for mosquitos. Now we try to prevent our children from going there. (Hamsici 2010, 188)

Duygu and others in Borçka mourn for the river as they linger by the corpse of it. But they mourn for more than that, for the lost opportunity to retrieve their childhood memories by looking at their children playing in the river.

5 Ethics, Ontology, Relationality

GRASSROOTS ENVIRONMENTALISM AND
THE NOTION OF SOCIO-ECOLOGICAL JUSTICE

"What is this injustice for? Should we all die here to protect our river?" cries a young woman from Loç Valley (WBR) against a private contractor who did not wait for the court decision to start digging up the bed of the local river to construct a hydropower plant, as depicted in the documentary film *Akıntıya Karşı* ("Against the Current"). As is well established in social movement studies, a sense of injustice is one of the primary motivations behind social and environmental struggles (Moore 1978; McAdam 1982; Turner and Killian 1987; Gamson 1992; Čapek 1993; Taylor 2000; Schlosberg 2007). These struggles are not only motivated by the experience and sense of injustice; they also disclose new dimensions of justice by articulating claims that "transgress the established grammar of normal justice" (Fraser 2009, 59). In this sense, justice cannot only be defined as a set of established formal principles and procedures;[1] it is a "permanent invention" (Balibar 2012), perpetually being expanded by the struggles arising on the threshold between procedural justice and its structural excesses (Balibar, Mezzadra, and Samaddar 2012).

Justice, thus, is always in excess of the law (Derrida 1992), and social struggles reveal the tension between the not-yet-captured, emergent dimensions of justice and historically given regimes of justice (see, e.g., Young 1990; Fraser 2009; Forst 2014). This tension is manifested in the concrete, day-to-day experiences of injustice and performed by the claims of those whose experiences of injustice are not represented by existing regimes of justice. It is, therefore, the movements that carry out what Kurasawa (2007) calls the "social labor of justice." Contemporary environmental movements, especially place-based, Indigenous, and local grassroots environmental struggles, have become ever more involved in the last few decades in undertaking this social labor of justice, that is, inventing new dimensions of justice that "transgress the established grammar" of existing justice regimes. In this manner, the prevalence of the concept of justice in contemporary environmental movements and scholarship is not at all coincidental.

In this chapter I examine the experiences of injustice and the claims of justice that contemporary environmental struggles build on from the body-centered, phenomenological perspective developed throughout the book. I employ a critical phenomenological perspective along with theories coming out of environmental humanities, Indigenous studies, and posthumanist studies to explore the relationality of human and nonhuman existence beyond culture and belief, and to discuss the conceptual implications of relational ontology and "grounded normativity" (Coulthard and Simpson 2016) for our understanding of justice. To achieve this aim, I take the empirical case of the antihydropower movement in Turkey and frame it as a "justice-seeking" movement, to use Samaddar's (2007) terminology, to expand the borders of existing theories and models of justice. In doing so, this chapter builds on the idea that the theory of justice can, and should, be rethought and reconfigured in light of emergent notions of justice arising from below, from within the movements and struggles and from the practices of their actors. Adopting an action-theoretical perspective and drawing on Michael Burawoy's extended case method (1998), I demonstrate how an empirical case study can be used to elaborate on broader questions of social and theoretical significance—in this case, of justice.

WHAT DOES GRASSROOTS ENVIRONMENTALISM HAVE TO
DO WITH JUSTICE? THE ENVIRONMENTAL JUSTICE
FRAME AND ITS LIMITS

Initially associated with the anti-toxic-waste struggles of Black communities in the United States in the early 1980s, environmental justice, both as
a movement and as a conceptual framework, has expanded to include a
broad range of grassroots environmental struggles in many different parts
of the world (Walker 2009; Schlosberg 2013; Martinez-Alier et al. 2016).
In the last few decades, environmental justice has left a mark in two main
ways. The first is the effective transformation of our perception of the
environment. The movements of local communities—rural, Indigenous,
Black, and minority communities all over the globe—against the immediate environmental threats that put their health, livelihoods, and lifeworlds
at risk have drastically altered the framing of the environment. The environment has come to denote the immediate environment, "where we live,
work and play" (Novotny 2000), instead of being somewhere out there "in
nature" that needs to be protected and conserved. Consequently, the environmental justice movement has fundamentally challenged the post-
materialist framing of environmentalism as a luxury issue (Inglehart
1990; Martinez-Alier 1995). Rather, the environment has been reframed
as a vital cause for the working classes, the rural populations, the poor, the
Indigenous, and the racialized and marginalized communities that have
been directly subjected to environmental hazards.

Another achievement of environmental justice, as a social movement
and as a body of scholarship, has been to yoke together the issues of environment and justice, mainly by revealing the indisputable correlation
between race, class, and gender, on the one hand, and the quality of the
environment we live in on the other. The "justice" implied in the articulation of the term *environmental justice* refers by and large to the (un)fair
distribution of environmental hazards and benefits. In this sense, the
environmental justice frame was initially understood as the application of
the Rawlsian idea of distributive justice to environmental issues.[2] This
theme of (un)fair and (un)just distribution has been extensively employed
in the environmental justice literature to frame the environmental grievances local communities suffer as an issue of (in)justice (Schlosberg

2007). As the environmental justice frame has expanded from narrowly focusing on the spatial distribution of waste and toxicity mainly in the United States to include diverse environmental issues in many different parts of the world, however, the idea of justice implied in the environmental justice frame has also expanded. Representation (or procedural justice) and recognition are encompassed first in the practices of claims-making and then in the definition of environmental justice (see Walker 2012; Agyeman et al. 2016).

Recognition and representation are being integrated more systematically into the environmental justice framework. Nancy Fraser's tripartite model of justice, in which redistribution, recognition, and representation correspond to the economic, cultural, and political realms of social justice, respectively (Fraser 2009, 15, 58–59), has become a major theoretical reference in the field (see, e.g., Schlosberg 2004, 2007; Ramos 2015; Coolsaet 2015; Svarstad and Benjaminsen 2020).[3] Fraser's model has provided environmental justice scholars with a useful conceptual framework for addressing the multiplicity and intersectionality of justice claims posed by the movements on the ground. My ethnographic research in the three regions of Turkey that are the focus of this book (the Mediterranean, Black Sea, and East and Southeast Anatolia [Kurdish] Regions) suggests that the aspects of redistribution, recognition, and representation intertwine differently in each case, depending on specificities of the region such as climate and the specific features of the natural landscape; the specific organization of socio-spatial relations and cultural identities; political heritage; the material properties of river waters; and the specific ways people relate to nonhuman life and the environment. I have already outlined in chapter 1 how the main motivations, dynamics, and discourses of the antihydropower struggle in each region are shaped by those geographical, cultural, and socio-ecological specificities, so I will not repeat that here. In this chapter, I want to explore whether and how the multiple notions of justice, as formulated by Fraser and employed widely in the environmental justice scholarship, help us decipher the claims of justice implied by those regional struggles.

Starting with redistribution, one should first recognize that the concept of (re)distribution, which concerns "fair allocation of divisible goods" (Fraser 2009, 3), might not be sufficient to capture the economic dimension of (in)justice implied by grassroots environmental struggles. Redistribution seems to relate to the idea of the welfare state and the state's role in distributing wealth among citizens through certain social schemes and subsidies. Here the condition is somewhat different. What is at stake here is losing what you already have, not as property but as commons, rather than not having a share of the wealth produced. It is, thus, more a case of dispossession than of maldistribution. The economic dimension of (in)justice, which is central to the "environmentalism of the poor" (Martinez-Alier 2002; Nixon 2011), is significant in the Turkish case, most notably in the Mediterranean Region, where the long and dry summers require irrigated farming, and rivers are the primary sources of irrigation for subsistence agriculture. In such cases, the struggle's primary motivation deals with the loss of the essential means of subsistence, of not being able to grow fruits and vegetables, and the fear of being forced to migrate to big cities only to join the army of the unemployed.

While the economic dimension of justice is at stake in the Mediterranean Region, claims for recognition are predominant in the Kurdish region,[4] where the entire issue is situated within the broader context of the Kurdish struggle for autonomy. In the context of environmental justice struggles, recognition is mainly discussed in relation to the claims of Indigenous and minority populations to cultural respect and self-determination (see, e.g., Castree 2004; Mascarenhas 2007; Vermeylen and Walker 2011; Whyte 2011; Martin et al. 2014; Ulloa 2017; for an important critique of recognition as a framework for Indigenous struggles of self-determination, see Coulthard 2014). The issues of recognition and assimilation are often central to the struggles for environmental commons and environmental justice, where a native minority is trying to protect their lands, forests, and waters against the settler-colonial capitalist rule. In these contexts, as in the Kurdish case, sovereignty over land and water is seen as essential for political autonomy and self-determination. In the case of hydropower, dams or smaller hydroelectric power plant projects are perceived as an element of the Turkish state's forced cultural assimilation of the Kurdish people. Thus, the aspect of recognition seems to represent the claims of justice implied by the movements in the Kurdish areas, such as in the cases in Ilisu and Dersim, regarding issues such as cultural value and identity, as well as self-determination and sovereignty.

Representation, on the other hand, "furnishes the stage on which struggles over distribution and recognition played out," as Fraser puts it (2009, 17). Indeed, one could posit (mis)representation as the underlying aspect in all three regions—Mediterranean, East and Southeast Anatolia (including the Kurdish areas), and East Black Sea—as the struggle, in all cases, is primarily caused by the fact that the people most affected by the decision to transfer the usage rights of water to private companies are not included in the decision-making processes concerning their environment. What Fraser calls "ordinary political misrepresentation" motivates the struggle,[5] as is made clear in narratives of being ignored, discounted, and excluded from decision-making, which are common in all cases (see chapter 1). Most locals I have interviewed told me that what hurts them the most is that they are discounted, "unimagined" as Rob Nixon (2011) would say, not treated as citizens or even as human beings. Meryem, a middle-aged woman from Kamilet Valley (EBR), expressed this succinctly: "Did they ask me if I want these HEPPs in my valley? They did not. Do I want them to intervene in my life like this? I do not. It hurts me that the state does not want to hear me, hear us." Her words convey more than her individual position against "ordinary misrepresentation"; she is voicing a collective sense of injustice that is echoed in the environmental struggles of local communities across rural Anatolia.

What these cases show us, then, is that Fraser's tripartite model of justice—setting aside the limitations of the concepts—can be usefully employed to decipher injustices that local communities in different regions face and to decode the claims of justice implied by their struggles against hydropower plants. However, I claim that the economic, cultural, and political conceptions, while necessary, are insufficient to translate the claims of justice produced within grassroots environmental struggles. Building on the analysis I have been developing throughout the book, I articulate the notion of *socio-ecological justice* in this chapter (see also Yaka 2019b, 2020), which lies at the core of the justice claims of the movements against hydropower plants, and potentially of other grassroots environmental struggles.

In the case of antihydropower struggles in Turkey, the river is the beating heart of the justice claims, not only as an economic resource that sustains livelihoods and as a cultural symbol of ethnic and religious identity or political autonomy, but also as a nonhuman entity, as a place maker, and as a relative (see below). Rivers shape the spatial imaginary, affective landscape, and symbolic order through their material properties and contingent behaviors as important *actants* (Latour 2004). This chapter demonstrates that local antihydropower movements in Turkey are not only against economic dispossession and cultural assimilation, but also for preserving a particular socio-ecological existence marked by the intrinsic relationship people establish with the river.

This framework became especially clear during my ethnographic research in the East Black Sea Region, the hotbed of resistance to hydropower plants. As I have already explained why anti-HEPP struggles cannot be adequately understood and studied using economic and cultural frames (see chapter 1), I will not repeat that discussion here. Nevertheless, a brief reminder is in order: rivers indeed bear cultural value in the region as in many other places, but in the broader sense of the term, as an essential part of the lifeworld, as an entity in relation to which "meaning and identities are produced" (Ahlers 2010, 224) through and within the habitual practices of everyday life. One clear example is the folk music of the region, in which rivers frequently feature as symbols of vitality, fertility, and joy and as the witnesses of life, love, and sorrow. Their cultural value, thus, is grounded in the everyday patterns of interaction between river waters and human communities more than in their symbolic significance for group identities and belief systems. Living in an East Black Sea village involves daily interaction with rivers. Life, in that sense, is a practice of living together and along with the river. In this sense, one might say that the cultural value of the rivers is socially grounded, built on the role of rivers in everyday social and affective life, rather than stemming from any symbolic status or sacredness.

The famous slogan of the anti-HEPP movement, *Su Hayattır* ("Water Is Life"), a version of which is used by many communities and movements across the globe, from Standing Rock in the United States to the Tsleil-Waututh Nation in Canada, from rural populations of South Africa to Indigenous communities of Guatemala, captures the centrality of rivers to life, which cannot be reduced to livelihood or belief. The multilayered interrelationship between water and life goes beyond these two established frameworks (of distribution/subsistence and recognition), as the case of the East Black Sea Region demonstrates. In other words, even though these two frames hold the conceptual potential to translate the justice claims produced within anti-HEPP struggles (particularly in the Mediterranean and Kurdish parts of the country), there remains an *excess of relationality* that is empirically manifested in the case of the East Black Sea Region. This excess of relationality over the established notions of environmental justice points to the relational and ecological constellation of human life in its entanglement with, and dependence on, water, land, and air, and a particular way of cohabiting with nonhuman beings within a web of life (see, e.g., Ingold 2012). The difference between this notion of relationality and economic notions of environmental justice is clear. My argument is that this difference is also not reducible to the cultural, to the frame of recognition, but goes beyond and is in excess of that frame as well.[6]

RELATIONAL ONTOLOGIES BEYOND RECOGNITION, CULTURE, AND BELIEF

> The feeling of strangeness that another culture provokes is of
> interest only if it leads one to reflect on the strangeness of
> one's own; otherwise it degenerates into exoticism,
> Orientalism, Occidentalism.
>
> Bruno Latour, *Politics of Nature: How to Bring the Sciences into Democracy*

"We are like the flesh and the fingernail" (*etle tırnak* in Turkish), said Elif, a young woman in her thirties, of her relationship to the river, as we spoke in front of her house by Arılı River (EBR). The saying refers to the indivisibility of two things, the impossibility of one being without the other, the principle of which is defined by Merleau-Ponty (1968, 208) as *being-in-indivision*. The principle of *being-in-indivision* captures the core aspect of relational ontology, carrying the conceptual ground further than *being-in-relation* (see Nancy 2000; Ollman 2003; Barad 2007; Benjamin 2015). Not only is our social and biological being necessarily being-in-relation, with these dynamic relations, instead of being subjects and objects, "form[ing] the basic material of reality" (Rosa 2019, 36), but no

entity is imaginable as separate from other (human and nonhuman) entities and their relations. In other words, as Muraca (2016, 19) states with
reference to A. N. Whitehead, "relations are ontologically prior to and constitutive of entities rather than being conceived as external link(ing)
between them."

Relational ontologies "that eschew the divisions between nature and
culture, individual and community, and between us and them" (Escobar
2011, 139) are increasingly being explored in their existing forms by
scholars of ecological and environmental anthropology, multispecies ethnography, and Indigenous studies (see, e.g., Descola 1996, 2013; Bird-
David 1999; Ingold 2000; Surrallés and Hierra 2005; Haraway 2008;
Lloyd et al. 2012; Datta 2015; Tsing 2015; Rosiek, Synder, and Pratt
2020). There is a tendency in the political ecology and environmental justice literature, however, to reduce relational ontological conceptions of
human and nonhuman life to culture, belief, and spirituality. Because of
this reduction of relational ontologies to culture and belief, Indigenous
environmental struggles are often discussed within the literature under
the conceptual banner of recognition. Relational ontological conceptions
of human and nonhuman existence, however, appear as a common, cross-
cultural pattern among many communities across the globe. Consequently,
it is not easy, analytically, to reduce relational ontologies to merely a cultural and/or spiritual element. Peoples from Amazonia to the Canadian
Arctic, from Siberia to Africa resist the modern/Western differentiation of
society and nature, living "within a continuum of interactions between
human and nonhuman persons" despite their differences (Descola
2013, 20).

The river as a nonhuman person is also a common theme that appears
among a variety of rural and Indigenous communities around the world.
A particularly widespread expression of this theme is seeing the river as a
relative. Metin, a young man and a local anti-HEPP activist from the village of Arılı (EBR), told me by the Arılı River that flows in front of his
house: "We grow up by this river. We are in contact with it every day; every
day, we see it, we hear it. It is like a neighbor, like a relative to us." Chas
Jewett, a Lakota woman and community leader, defines the Missouri
River similarly: "It's a relative that nourishes us" (Jewett and Garavan
2019, 44). Indigenous scholar Glen Coulthard and his colleague Kate

Neville quote Lakota historian Nick Estes, who maintains that Indigenous resistance "to trespass[ing] of settlers, pipelines and dams, is part of being a good relative [*Wotakuye* in the Lakota language] to the water, land and animals" (Estes quoted in Neville and Coulthard 2019, 2). A relative could be a brother, as Leni, a young Awajun leader of the Indigenous environmental struggle in the Amazonian lowlands of northern Peru explains: "We speak of our brothers who quench our thirst, who bathe us, those who protect our needs—this [brother] is what we call the river. We do not use the river for our sewage; a brother cannot stab another brother. We do not stab our brothers" (Leni in De la Cadena 2010, 363). A river could also be a child, as Salih Gut, a young man from the village of Çiftlik in Korgan District (EBR), articulates: "The river is an orphan child. A child of all villagers and s/he[7] is about to die. S/he screams, s/he says, "Save me!" The villagers are trying to save this orphan child—we will save her/him. We won't let this child be fed to rent and profit."[8] Another example is given by Muhammet Kaçar (2017), who reports in the daily newspaper *Hürriyet* about a group of women who blocked the road in the Arılı valley and who talked to the experts appointed by the court for the HEPP case. One of them told him, *"Dere bizim eşimiz, biz derenin eşiyiz"*—"The river is our spouse; we are the river's spouse." The word *eş* in Turkish is used for partner or spouse in everyday life, but it also has the connotation of being one's equal, fellow, companion. *Dere* (small river/stream) can be all of them—a partner, a relative, a brother, a sister, a friend, a neighbor, a companion, all in all an equal—a nonhuman person/being/entity with whom one has a lifelong, intimate, sentient, and affective relationship.

Based on their case study of the water ontologies of four Yukon First Nations, Wilson and Inkster (2018) define this relationship as a type of kinship. More-than-human kinship, which has become a key theme in the posthumanist literature through the work of Donna Haraway, seems to capture the nature of the relationship human communities establish with the rivers they live with. Indeed, this notion of the river as a relative, a kin, speaks to a growing body of literature within multiple fields—such as science and technology studies; posthumanist and new materialist theories; critical, nonrepresentational, and postphenomenological geographies; and environmental humanities—that emphasize that what we define as "social" is entangled with and dependent on the nonhuman world (see, e.g., Haraway 1991; Latour 1993; Swyngedouw 1999; Ingold 2000; Milton 2002; Whatmore 2002; Thrift 2008; Alaimo 2010; Bennett 2010; Braidotti 2013; Descola 2013; Ash and Simpson 2016; Whitmore 2018; Simonsen and Koefoed 2020).

What is ignored by the canonical social and environmental justice theories is precisely this dimension of sociality: that the social world involves nonhuman entities and environments. The social world is thus "a more than human one" (Simonsen and Koefoed 2020, 14). This is why nonhuman animals, and sometimes environmental entities such as rivers, are seen as nonhuman persons. It does not mean, of course, that human and nonhuman persons are all the same. They stand on a spectrum of differences in terms of perceptual and cognitive capacities and dispositions. But these are differences within a relational existence, as we all exist in a corporeal continuity with other bodies and things. What is articulated as animism in modern anthropology, that is, the personification of nonhuman animals and things, does not ignore the differences; it rather acknowledges the relationality and marks a sociality, an intersubjective field that involves nonhuman beings. As Bird-David (1999) states, personification is very much about being in relation and in conversation—we personify nonhuman beings as we socialize with them. Animism, in this sense, is engendered by "non-representational, affective interactions with other-than-humans" (de la Cadena 2010, 346). It is a manifestation of "an interactive, participatory exchange—a kind of non-verbal conversation— with the things that surround" (Abram 2017, 278).

Understanding animism as an aspect of "responsive relatedness," which "amounts to a kind of sensory participation" (Ingold 1999, 82), and drawing on the work of Merleau-Ponty, one could maintain that animism is not peculiar to particular cultures, but is inherent in the act of perception. In other words, "sensorial perception is inherently animistic" and "participatory—involves the experience of an active interplay or coupling between the perceiving body and that which it perceives" (Abram 2017, 57, 277–8). Our bodily senses and affects are the media of our connectivity with this more-than-human world, and, thus, play a crucial role in the emergence of relational ontologies. A social and affective relation is built through sensory perception, through the embodied and sensory connection of human and nonhuman bodies, entities, and intensities. This social and affective relation between human and nonhuman beings—a relation that is conditioned by corporeal encounters that occur within the common web of "flesh" that we share with nonhuman bodies, entities, and organisms (Merleau-Ponty 1968, 2003; see also chapter 3)—engenders identification and empathy, that is, to invoke the phenomenological language, *Einfühlung*.

Besides *Einfühlung*, Merleau-Ponty uses another German term, taken also from Husserl, to discuss the aspect of relationality in his course notes on nature: *Ineinander*. He defines *Ineinander* (literally, "into-one-another") as "the inherence of the self in the world, and of the world in the self, of the self in the other and the other in the self" (Merleau-Ponty 2003, 306). What the term communicates, in this sense, is close to the contemporary understanding of an ecological, expanded, and relational self "that functions in a nature-culture continuum" (Braidotti 2013, 61; see also Fox 1986, 1990; Naess 1989; Bird-David 1999). Thinking *Einfühlung* and *Ineinander* together, we can postulate that the expanded, relational, and/or ecological self, an understanding of the self that incorporates its relations with human and nonhuman others (*Ineinander*), is based on this experience of identification and empathy (*Einfühlung*).

It might be true that identifying with nonhuman life and the consequent expansion of the self to include nonhuman beings and entities seems to be observed mainly in Indigenous and rural communities (see Posey 1999). It does not necessarily follow, however, that the relational existence of the self as transversally connected to and interdependent with the nonhuman life is peculiar to Indigenous cultures or that it is to be attributed to culture and/or cosmology as a "corpus of traditional wisdom," as Ingold would put it (2000, 137). It is, instead, the human condition. The accelerated, isolated, increasingly mobile and mediated structures of the modern/Western lifeworld disguise the ecological embeddedness of our social existence, creating an illusion of power, autonomy, and independence. It becomes increasingly difficult to attune oneself to the nonverbal conversation, which is inherent in sensory perception, within the routine, habitual ways in which we live our lives (see Rosa 2019 on the concept of resonance; see also Abram 2017). It is for this reason that we are more likely to come across explicit expressions of relational ontological understandings of human-nonhuman existence with regard to place-based communities that live in close engagement and intimate sensory-affective connection with their environments across the globe. In other words, a certain way of experiencing and relating to the more-than-human lifeworld, a close and continuous engagement with the nonhuman environment, engenders an intimate affective and sensory relationship with it and, thus, facilitates a relational understanding of human and nonhuman life (see, e.g., Ingold 1999, 2000; Descola 1996, 2013; Strang 2005). Hence it is not an essential, qualitative difference that engenders relational ontologies; it is a difference of experience—of how we experience and relate to the world.

This difference could well be depicted as cultural, spiritual, belief-related, and cosmological. And indeed, many Indigenous cultures, cosmologies, and beliefs involve a relational ontological understanding of human and nonhuman existence. My intention here is not to ignore or deny their existence and efficacy (see Gergan 2015). It is rather to object to the *reduction* and *confinement* of relational ontologies to culture and belief, which practically strips them of their ontological potential to transform hegemonic conceptions of society and nature as dualistic categories. De la Cadena (2010, 346) discusses the reduction and containment of relational ontologies and "earth practices" either to belief and spirituality or to "cultural interpretations" of nature "worthy of preservation as long as they did not claim their right to define reality."[9] She maintains that the "relational condition between human and other-than-human beings," which defies "the dominant ontological distinction between humans and nature" (de la Cadena 2010, 341) and constitutes the lifeworld of people (in the Andes and in many other places), is a matter of historical and political ontology beyond culture and belief. Chas Jewett, the Lakota woman I cited above, also makes a similar point when she maintains that the Lakota relationship with the Earth—"being related to everything—air, water, trees"—is not a question of religion, belief, or spirituality in the Western sense; rather, it is who they are, their mode of being (Jewett and Garavan 2019, 50). Drawing on these accounts, it becomes possible to imagine another way of conceptualizing the "already existing" (Thomas 2015) relational ontologies—whether they are elements of a specific cosmological vision or not. From a phenomenological perspective, we can conceive of them as ontological articulations of our reciprocal, participatory interchange with the nonhuman environment, produced within the everyday bodily practices that involve constant sensory and affective interaction with nonhuman entities. Hence, rather than reducing them to culture and belief, it is possible to understand existing forms of relational ontological understandings as manifestations of the ecological embeddedness of human life in general.

De la Cadena and Jewett's remarks also bring to mind Tim Ingold's seminal *The Perception of the Environment*, in which he writes that "what the anthropologist calls cosmology is, for the people themselves, a lifeworld" (Ingold 2000, 14). It could be helpful here to understand cosmology and belief not as a "separate and immutable feature of the world" but as "embedded in everyday life" (Whitmore 2018, 27), rooted in the embodied, material, sensory, and affective relations with the human and nonhuman world in which we live.[10] Based on her research on the Yavapai people who fought against the Orme Dam in Arizona, Espeland (1998, 200) writes, "Beliefs about land are not abstract ideas but relationships that are enacted daily on the reservation when residents swim in the river, wander the desert, watch the eagles, or wake up to the sight of the mountains." Drew (2017) also discusses the same phenomenon based on the everyday interactions of the Garhwali women with the Ganga (the Ganges River), in the context of antidam struggles in the region:

> To better understand why large dams on the Ganga would upset people, such as the women living along its flow in the mountains, it is necessary to appreciate the river's role in daily practice and everyday life. While these moments of encounter are shaped by notions of the Ganga's significance in various Hindu texts and religious teachings, they are also phenomenological points of connection that regularly verify the goddess's powers and her blessings to devotees. (Drew 2017, 48)

Drawing on their ethnographic work, Whitmore, Espeland, and Drew maintain that everyday relationships and bodily interactions are central to the enactment of the cosmological and/or religious value of natural entities such as the land and water.[11]

FROM RELATIONAL ONTOLOGIES TO RELATIONAL ETHICS

> Can you imagine a world where nature is understood as full
> of relatives not resources, where inalienable rights are
> balanced with inalienable responsibilities wealth itself is
> measured not by resource ownership and control, but by the
> number of good relationships we maintain in the complex
> and diverse life-systems of this blue green planet? I can.
>
> Daniel Wildcat, quoted by Soren C. Larsen and Jay T. Johnson
> in *Being Together in Place: Indigenous Coexistence in a More-than-
> Human World*

Everyday relationships and bodily interactions with nonhuman entities engender a social and affective relation. It is this social and affective relation that effectively "enlarges the boundary of the community to include soils, waters, plants and animals" (Dobson 2000, 42), what Leopold (1949) once coined a "land ethic." Such an ethical understanding lies beneath the surface in many grassroots environmental struggles in which local communities fight not only for their isolated, immediate "interests," as the social movement scholars inspired by rational choice theory might expect, but to defend a certain way of coexistence in which animals, waters, forests, and lands are seen not just as resources, but as companions (see Escobar 2011). If such movements must be defined within the frame of interests, it should not be an isolated and narrowly defined frame, but an expanded notion of interests in terms of acknowledging our situatedness in and dependence on nonhuman ecologies, building on the awareness that destroying them is a form of "auto-destruction" (Guattari 2000; see also Porritt 1984; Fox 1986; Dobson 2000; Bennett 2010).

This expanded notion of interests is closely related to the expanded (or relational) self, discussed above, an understanding of the self that incorporates its relations with human and nonhuman others. Bird-David (1999) reminds us that the term *dividual* is derived from ethnographic work and designates an understanding of the person as constituted by relationships, in contrast to the modern/Western conception of the *individual*, defined as a single, independent entity. It is this understanding of *dividual*, of an expansive self and the tendency to relate to and identify with the nonhuman world, that is central to what we call relational ethics—an ethical understanding that builds on reciprocity, respect, and care for nonhuman others.

Scholars of ecological ethics and environmental humanities, such as Warwick Fox (1986, 1990) and Kay Milton (2002), emphasize that ecological consciousness often stems from our deep identification with the nonhuman world rather than from complex ethical considerations of its intrinsic value (see also Abram 2017). Ethics, thus, could also be cultivated as a "state of being" rather than as a "code of conduct," drawing on our relational and transpersonal experience of the world (Dobson 2000). Personification is an aspect of this relational and transpersonal experience that results in ethical codes of conduct. Descola (2013) writes about the Achuar hunters' ethic, which involves understanding the animals they kill to eat as persons worthy of respect and forbids killing more than one needs.[12] Yakup Okumuşoğlu, a well-known lawyer who has represented on a pro bono basis many local communities across the East Black Sea Region in their legal fights against HEPPs, mentions a similar code of ethical conduct in an interview with Sinan Erensü. Okumuşoğlu, who grew up in the region himself, explains the local ethics of forestry as he experienced it as a child in the district of Çamlıhemşin (EBR):

> My grandfather used to cover the blade of his axe with a piece of cloth as he went to the forest to cut wood for the winter. It means that the forest is alive; it could understand that we were there to cut its trees and be sad. Blades must be covered to prevent that. . . . One should not hurt the trees, should not cut the ones which are still alive, only the ones that are already dead [dried out]. (Erensü 2016b, 437)

Such an understanding of ethics that builds on our relational experience of the world is manifested in the "already existing relational ethics" (Thomas 2015, 978) in many Indigenous and place-based communities in the form of responsibilities and obligations toward nonhuman entities such as the land, animals, rivers, lakes, and trees (see Jewett and Garavan 2019; Coulthard 2010).[13] Indigenous scholars Glen Coulthard and Leanne Betasamosake Simpson (2016) call this practice-based relational ethics *grounded normativity*. What is common to different examples of Indigenous relational ethics regarding the rivers and river waters is perceiving and respecting the river as a sentient, living being beyond the conventional modern/Western perception of water as a resource (see, e.g., McGregor 2015; Thomas 2015; Yates, Harris, and Wilson 2017; Wilson and Inkster 2018; Bannon 2020). Thanks to struggles of Indigenous communities, this conception of rivers and other nonhuman environmental entities such as mountains and forests as living beings has been permeating the legal sphere and transforming our understandings of justice. The recognition of the Whanganui and Atrato Rivers as legal persons, in New Zealand and Columbia respectively, as well as incorporation of nature's rights into the constitutions of Ecuador and Bolivia, are examples of the infiltration of relational ethics to expand the borders of the existing regimes of justice by altering legal frameworks.

Yates, Harris, and Wilson (2017) depict the ontological difference in the ways water is perceived, used, and experienced by contrasting modern ontologies of water-as-resource and Indigenous/relational ontologies of water-as-lifeblood, the latter defined as "flow oriented and connecting." In the case of the East Black Sea Region, we can see the elements of just such a relational ontological-cum-ethical perspective, such as the common use of the *water-as-lifeblood* terminology (see chapter 3) and the respect given to water as a living and life-giving entity, associating the circulation of blood and the flow of the river as equally fundamental to (human and non-human) life. Material properties of water are important here, as "the flow of water between bodies and environments . . . leads to deeply relational ideas about common substance and connection" (Strang 2014, 138). This common substance and corporeal continuity between human body and nonhuman bodies, organisms, and things, also depicted by Merleau-Ponty with the concept of flesh, is certainly an important reference point for conceptions and practices of relational environmental ethics.

In the posthumanist and new materialist circles, there is a tendency to assume that corporeal continuity and/or kinship, the fact that we are "folds of world's flesh" (Toadvine 2009, 134) as Merleau-Ponty would put it, has direct ethical, and sometimes also political, implications (see, e.g., Whatmore 2002; Alaimo 2010; Bennett 2010). The phenomenological position, on the other hand, affirms corporeal continuity as a condition of possibility for an intimate relationship with the nonhuman world in that it "makes room for affective, habitual and visceral dimensions of body relationships, inorganic as well as organic, and broadens the scope of ethics to include such relationships" (Toadvine 2015, 182). But our common materiality does not directly determine our ethical conduct (Toadvine 2009, 2015). What would transform our ethical conduct toward the nonhuman world is the mode of our relations with it (whether such an intimate, corporeal relation is shaping our lifeworld); it is the way we experience and relate to it through our bodily capacities and dispositions.

In the case of water, even though our corporeal continuity with water bodies—the fact that we are water bodies ourselves (Neimanis 2017)—makes an impact on our relationship in terms of facilitating certain sensory and affective experiences (of flowing and floating), what engenders relational ethical practices in the case of Indigenous communities, and in the case of the East Black Sea Region, is a certain way of relating to and living with water bodies. It is through these ways of relating to and living with that water surpasses its instrumental role in sustaining biological life on Earth and becomes a part of social relations (Krause and Strang 2016). And we personify and identify with the rivers not because of our common substance, but because we develop a social and affective relationship with them, a relationship that is transferred from one generation to the next (see chapter 4).

DEVELOPING THE NOTION OF SOCIO-ECOLOGICAL JUSTICE

The notion of socio-ecological justice builds on the conception of relationality (of human and nonhuman life, of the social and the ecological, of nature and culture), drawing on the existing practices of relational ontologies and ethics. It employs eco-phenomenology, green theory, ecological anthropology, multispecies ethnography, and Indigenous studies on the one hand, and posthumanist, new materialist, nonrepresentational, actor-network theories on the other, to break with the society-nature dualism within the field of justice. It is well established that the processes of modernization maintained nature as external to human life and society, as the realm of necessity to be emancipated from, and as wilderness to be dominated. Along similar lines, nonhuman life and nature are treated merely as resources in the service of humanity. We became accustomed to thinking of the social and the ecological as strictly separate realms: ecology is the interrelationships between organisms, while society is about the interrelationships between human beings.

It is firmly maintained in the fields of ecological anthropology, environmental humanities, and Indigenous studies, however, that understanding nature as separate from and external to the social and cultural spheres is by no means universal. To illustrate with an example from our case of the East Black Sea Region: Nihan Bozok, Mehmet Bozok, and Meral Akbaş (2016) write that nature as a concept made no sense to the older women they interviewed in the remote villages of the region (bordering Georgia). These women were able to talk about their relationship with the river, with the sea, with the rain, with trees, and with the plants and animals that surround them. But nature as an ontological category, distinct from society and culture, made no sense to them as those rivers, trees, plants, and animals they live with were an essential part of their social and cultural life. Nature here is not something separate from the community; it is instead "a part of the social world" (Gudynas 2011).

Nature-society dualism is also proving to be an increasingly inadequate frame of reference for facing the current challenges of the climate crisis, as an acute awareness of "transversal" interconnection is imposed on us. We come to realize, on the verge of an ecological crisis, that relationships between humans cannot be imagined without the mediation of nonhuman entities. And we are becoming increasingly aware, as Latour puts it, that "nature and society do not designate domains of reality; instead, they refer to a quite specific form of public organization" (Latour 2004, 53). In other words, we are currently facing the fact that the ecological embeddedness of human life characterizes not only the local and Indigenous communities of remote regions; all of human life is ecologically embedded, in different ways in every temporal and spatial context, albeit temporarily overshadowed by the modern organization of the nature-society duality. This modern organization of the nature-society duality, however, seems to be squeezed between the "premodern" arbitrariness of its very existence and the ultramodern blurring of it by biotechnology and biomedicine, genetics, and epidemiology, hybrids and cyborgs, and the like. Indeed, it is becoming increasingly difficult to deny that human life, sociality, and agency are increasingly entangled with and dependent on nonhuman ecologies (see Haraway 1991; Latour 1993; Bennett 2010).

One needs to stress, though, that not only the biological body (Alaimo 2010; Bennett 2010), but also self and subjectivity are assemblages of relations that involve nonhuman bodies, organisms, and objects (Bird-David 1999; Bennett 2010; Braidotti 2013). We sense our bodies only in other (human and nonhuman) bodies and things. To put it in phenomenological terms, it is not only that we become part of the flesh of the world through our own flesh, but we also sense the flesh of our own body only within the flesh of the world. We sense, we feel, we act, we come to know ourselves only through our environments, through our connectedness with other bodies, organisms, and things. The self, in this sense, is formed in relation to the other, within a world of encounters, not only with human but also with nonhuman bodies and entities. In other words, everyday relations, interactions, and encounters that form selfhood and subjectivity involve both human and nonhuman others.

From such a relational perspective, human life, agency, and sociality exist only in relation to nonhuman ecologies, and the relationship between them is not external to, but co-constitutive of, their existence. In that sense, as Ingold (2000) argues, the intrahuman relations that we are "accustomed to call social" should be seen as a subset of ecological relations. Or, to put it differently, our social world is a more-than-human world (see, e.g., Haraway 1991; Law and Mol 1995; Whatmore 2002; Braun 2004; Latour 2007; Whitmore 2018). As human societies are in an "intricate interdependency" and "transversal interconnection" with their nonhuman environments (Franklin et al. 2000; Braidotti 2013), what we call social is, ultimately, socio-ecological. The wording *socio-ecological*, hence, implies the need to rethink the idea of the social and sociality in relation to nonhuman ecologies—properties, entities, organisms, things, actants, and the like—and materialities, as "interactive practices" take place not only between human beings, but within a "more-than-human" world.

If the social can only be thought of in an intrinsic relation to the ecological, then our ideas of social justice should be rethought accordingly. Ethics and justice are often understood as a set of notions and principles concerned with intrahuman relations. There are of course attempts to extend their boundaries to include the nonhuman world, such as the concept of ecological justice (Low and Gleeson 1998; Baxter 2005). The concept of socio-ecological justice that I aim to introduce here, however, differs from the ecological justice that builds on discussions of the moral status of nonhuman species as subjects of justice. Instead, I understand socio-ecological justice as an aspect of social justice that signifies the interconnection between the so-called social and ecological realms. Socio-ecological justice maintains that our intrinsic and intimate relations with the nonhuman world are an essential part of our well-being and are central to our demands to pursue a fair, decent life. Thus, what is at stake here is not merely a matter of extending the community of justice to include nonhuman environments, but also of actively incorporating human-nonhuman relationality and conceptions and practices of relational ontologies and ethics into our understanding of justice (Yaka 2019b).

Socio-ecological justice is in excess of environmental justice as it is defined within the framework of justice as redistribution, recognition, and representation. The concept of environmental justice has been significant in relating environmental issues to power, justice, and inequality, and challenging the mainstream environmentalisms that had failed to recognize environmental inequalities within society (Mohai, Pellow, and Roberts 2009). The literature, however, mostly treats the environment as yet another setting that reflects the relations of domination and oppression embedded in society, such as racism, poverty, and patriarchy. Even though it is extremely important to point out these relations in each and every setting, the stress on distributional (in)justice of environmental hazards falls short of understanding justice as a broader issue of interdependence between "human and extra-human natures in the web of life" (Moore 2015, 5).[14] Socio-ecological justice could thus be seen as an attempt to go beyond the duality of ecological justice vs. environmental justice—either protecting nature from human society (rights of nature) or protecting humans from environmental hazards and deprivation (rights of humans regarding the environment)—which corresponds to the ontological distinction between human and nonhuman life. Instead, it locates justice within a relational ontology that maintains an intrinsic relationship between human and nonhuman existence, between social and ecological phenomena.

Such a notion of socio-ecological justice frames rights and interests of "humans-in-nature" in relation to the rights and interests of nonhuman nature, instead of the latter being in existential conflict with the former, as the ecological justice theory implies (see, e.g., Baxter 2005; Parris et al. 2014). As the case of the East Black Sea Region and many others demonstrate (see, e.g., Fox 1990; Milton 2002; Strang 2005), caring for one's own life and caring for the environment can be one and the same, so long as communities perceive themselves as part and parcel of the nonhuman world surrounding them. In the case of the struggles to protect rivers, for instance, local communities fight both for their own rights and for the rights of the river and other beings, as the river does not only belong to the people. It belongs to itself (Indigenous writer Lee Maracle, quoted in De Wolff and Faletti 2022, 7) and to all living things. It belongs to the fish swimming in it, to the tree roots that it hydrates, to the wild animals drinking from it, even to the rocks over which it flows—even though they would not be counted as living things in the classical sense. This is why villagers often mention other forms of life, such as the fish and the wild animals living in and by river waters. Vedat, for instance, a middle-aged man from the village of Balıklı (EBR), took me to his village to show me the river and asked, "What would happen to the fish? Would they survive in those pipes?" While Vedat mentioned the fish, Ayşegül, a young woman from Konaklı (EBR), talked about wild animals: "What about the bears, the deer, all those wild animals who drink from the river?"

Drawing on such empirical cases and theoretical sources, socio-ecological justice frames the relational existence of human and nonhuman ecologies as a matter of justice. To put it differently, it is an attempt to translate the relationality of the social and the ecological, of human life and the nonhuman world, into the vocabulary of justice. We need a new vocabulary, not only to dismantle the society-nature binary (Castree 2005), but to rearticulate concepts such as sociality, justice, subjectivity, even democracy, in order to facilitate a "progressive composition of a common world" (Latour 2004) with human and nonhuman others. Such a rearticulation entails reframing the social and the political as processes and struggles of our common life on Earth, not only with human but also with nonhuman others. In the words of Escobar (2011, 139), "[W]e need to stop bothering the Earth with the dualisms of the past centuries, and acknowledge the radical interrelatedness, open-ness and plurality that inhabit it." The current historical conjuncture requires an urgent recognition of our ecological situatedness within and dependence upon the ecosystems and the nonhuman world. It means dropping the illusions of autonomy, power, and independence with respect to the more-than-human world we live in, accepting our embeddedness, interdependence, and vulnerability (see Larsen and Johnson 2017). Socio-ecological justice builds on this acceptance; it recognizes and responds to the centrality of human-nonhuman relations, at least for human life on Earth, which has been phenomenologically overshadowed by the experience of modernity.

If we return to Fraser's terminology of redistribution, recognition, and representation, socio-ecological justice corresponds to the (right of) *coexistence* of human societies and nonhuman ecologies in its complex, internal, and multidimensional relationalities within a common planetary ecological system. Coexistence involves rights of nonhuman life and rights of humans-in-nature to live and flourish together in their environments, free from institutionally sustained destruction, degradation, pollution, toxification, and commodification of ecological systems, habitats, and entities. Drawing on the intrinsic relationship between ecological and social spheres, socio-ecological justice frames the notion of justice neither as strictly humanist nor as ecocentric. Instead, it puts the accent on the *relationality* of the human and nonhuman natures and intends to rethink justice on the basis of the relationship between them. Consequently, it aims to extend the borders of justice as a requirement to rethink social justice in a more-than-human world.

If justice is a "permanent invention" (Balibar 2012, 29), new dimensions of which are disclosed "through the medium of social struggle" (Fraser 2009, 59), then contemporary grassroots environmentalisms, which are, at their core, struggles against the injustices of environmental dispossession, are inventing justice anew by extending the borders of the established regimes of justice. They disclose a certain understanding of nonhuman entities that goes far beyond our conventional understandings of them as "resources." Those established conceptions take no conceptual notice of the fact that what we experience and identify as injustice and what we demand as justice necessarily involve our transversal connection to our environments. Socio-ecological justice aims to incorporate this connection into our conception of social justice by stressing the rights of human and nonhuman worlds to thrive and prosper in a "web of intricate interdependency." Locating justice within a relational ontology that maintains an intrinsic relationship between social and ecological phenomena and relating it to a more-than-human ethic, socio-ecological justice is a conceptual move toward a theory of justice that speaks to the posthuman condition and the Anthropocene (Braidotti 2013).

Conclusion

TOWARD AN ECOLOGICAL APPROACH TO
LIFEWORLD, SOCIALITY, AND AGENCY

Hydro dams, once celebrated as temples of the modern era (Nixon 2011), have been losing their prestige in the face of strong transnational campaigns that successfully publicize the socio-ecological destruction caused by dams even when they fail to hinder the actual construction processes. Their success has led corporate giants such as Disney to adopt the anti-dam position. Disney wrapped the message up in postcolonial discourse in its big hit franchise *Frozen*. As the parents of small children like myself would know, *Frozen 2* is about the white rulers of Arendelle building a dam on the land of the Indigenous Northuldra people, which poisons their land and imprisons the forest, together with the Northuldra people, within a heavy, magical fog. The dam is destroyed in the end, as an act of reparation by the Arendelle princess, Anna, for whom the clever scriptwriters of Disney invented a Northuldra mother to avoid the conventional white savior plotline.

Some hydro dams are indeed being dismantled as a result of the environmental struggles of Indigenous and rural communities; for instance, removal of two large dams on the Elwha River in the US state of Washington aimed to restore the river ecosystem and allow salmon to reach their spawning grounds. At the same time, however, small-scale run-of-the-river HEPPs are proliferating globally as the presumably eco-friendly alternative to large-scale hydro dams, despite the emerging evidence of their ecological impacts (Couto and Olden 2018). The initial aim of this book was to explore these seriously underestimated socio-ecological impacts, as they motivate countless local communities across Anatolia to oppose private, run-of-the-river HEPPs. This strong grass-roots resistance against extensive small-scale HEPP development in Turkey came as a surprise, even for environmental activists themselves. Prior to the anti-HEPP movement, Turkey had had a negligible history of rural, environmentally motivated resistance, the most well known being the peasants in Bergama, located in the Aegean Region, who fought against cyanide-based gold mining with substantial public support in the 1990s and early 2000s (Çoban 2004).[1] Anti-HEPP movements, however, have been unprecedented in their scale and scope, as the conflicts and protests have emerged in many different villages and valleys around hundreds of projects throughout Turkey.

At the time of this writing in late 2022, the struggle against HEPPs continues in Turkey, albeit at a less intense level. While many HEPP projects have been completed in the last two decades, many others have been blocked by local communities. In addition, the worsening economic situation in Turkey has forced energy and construction companies to reconsider their investments, including HEPP projects, in the last few years. In the meantime, it has been revealed that many HEPP projects have not proven to be as profitable as their initial projections promised, and sometimes were not even economically viable to begin with, as most of the small- and medium-sized streams on which hundreds of HEPPs were planned on paper do not have the water volume or energy potential necessary to sustain a HEPP or render the investment profitable. Still, widespread and efficacious local resistance has been the main reason investments in HEPPs have slowed down, as HEPP projects are now saddled with a bad reputation in the public sphere.[2] This unmatched prevalence and success of anti-HEPP movements clearly shows something about the urgency of the cause: river waters.

Interpreting this urgency, one immediately thinks about direct uses of water for subsistence, especially in agriculture and animal husbandry, in line with the livelihood argument on which the entire literature on "resource conflicts" is built. And sustaining livelihoods is indeed the primary motivation behind local resistance against HEPPs in certain parts of the country, such as in the Mediterranean Region, where summers are hot and dry and, consequently, irrigated agriculture is the norm. But this is not an overarching argument that can be applied equally to all regions of Turkey; it does not, for example, explain the resistance in the East Black Sea Region and the Kurdish region. Explaining the resistance in the Kurdish region is easier, as it is very much connected with the broader struggle for political recognition and Kurdish autonomy in the region. Explaining the local movements in the East Black Sea Region, however, especially in the coastal part that is most intensively targeted by countless HEPP projects, requires closer attention.

Why have these coastal (northern) parts of the East Black Sea Region, in which river waters are not used for agriculture or for any other immediate economic purpose, become so closely associated with the anti-HEPP struggle in Turkey? Why is this region the home of so many local movements, including symbolic ones, as in the case of Fındıklı? These questions have honed the focus of the book, as I discuss in the introduction. As I observed the radical activism of women against HEPPs in the East Black Sea Region and beyond, as well as the difference between the ways in which men and women communicate their grievances and motivations and express their political agencies, the focus of the book became twofold. These two aspects, however, are very much related, as I have shown throughout the book. The conceptual effort to explore the value of river waters beyond their immediate use as a "resource" (or beyond their symbolic cultural or cosmological value) goes hand in hand with the effort to understand women's environmental activism beyond the goal of sustaining livelihoods; the two efforts feed into and support one another in many different ways.

At the center of this conceptual effort stands the body and the corporeal relationship between human bodies and bodies of water (in this case, rivers). Drawing on the work of Merleau-Ponty, as well as on the feminist scholars who engage with his work, I have developed a body-centered approach throughout the book, taking the lived body as a key methodological tool for exploring the place-based, gendered, grassroots activism against HEPPs in Turkey in multiple ways. I employ the body, first and foremost, as "our general means of having a world" (Merleau-Ponty 2012, 147). It is the site of spatiality and the nexus of movement and experience through which we inhabit the world. One relates to (and with) the world through corporeal involvement and experience; one becomes conscious of herself and her environment through bodily senses and affects. The body as I operationalize it is not just the physical body "that I see, touch or sense; rather, it is my capacity to see, touch, and sense. It is not an object in the world, but the medium, the field, or the capacity that reveals the world to me" (Fuchs 2018, 73). The body, hence, is not a shell in which the ghost, or the spirit, or the mind, is located; it is formative, generative, and agential; it is the "very stuff of subjectivity" (Grosz 1994, ix).

Second, I deploy the body both as our anchorage in a world and our point of view upon that world (Merleau-Ponty 2012, 73, 146). We do not view "everything from nowhere" like a disembodied spirit, like God (Haraway 1988; see also Nagel 1986); we perceive the world and act upon it as embodied beings. Perceiving requires embodiment, and being embodied means being situated. We are situated in this material world that is shaped by multilayered socio-ecological relations of power and difference through our bodies in sexually, spatially, historically, and culturally specific ways. Body, thus, is the starting point for a "feminist politics of location," as Adrienne Rich (1984) once put it, as we—our perceptions and perspectives, knowledge and consciousness, subjectivities and identities—are located, first and foremost, in our bodies.

Third, I maintain that the body "is made of the same flesh as the world, . . . is the prolongation of the world" (Merleau-Ponty 1968, 248, 255). We are interconnected with the material world we inhabit through our material bodies, through our flesh, which is a part of the flesh of the world. Our body "unites us with the things through its ontogenesis" (Merleau-Ponty 1968, 136). Our bodily, material interconnection enables us to feel with (or *feel as one with*) the nonhuman bodies we encounter, that is, *Einfühlung*. In this manner, the body is also a methodological medium that allows us to realize Husserl's famous central principle of phenomenology: "to go back to the things themselves," or, to use a more postmodern formulation, to focus on the nonhuman. One of the main objectives of the approaches that led to the so-called material turn, such as actor-network theory, nonrepresentational theory, and new materialism, is to focus the attention on those practical engagements and bodily connections that involve nonhuman entities as well as human beings. In this sense, the material turn is mostly about bringing these "backgrounds" to the fore in the analysis of the social, the subject, and the world (Anderson and Harrison 2010). What phenomenology (especially the work of Merleau-Ponty and critical and feminist scholars who engage with his work) offers within this material turn is a methodology for bringing these backgrounds to the center of the analysis, not by erasing the first-person experience, but by tracing the bodily nature of experience to "use our bodies to understand how nonhuman objects shape human situations" (Ash 2017, 206).

FEMINIST PHENOMENOLOGY, RELATIONAL ONTOLOGY, AND THE ETHICS OF COEXISTENCE

Our connection to the nonhuman world is not just organic and material, but experiential. Phenomenological methodology maintains that this connection—that is, our habitual, practical engagements with the nonhuman world and the tacit, sentient, and affective aspects that those engagements routinely involve—is to be explored in order to "go back to the things themselves." These habitual engagements and affective connections are often ignored by mainstream analyses of human action and interaction. But, as the approaches I mentioned above have repeatedly demonstrated, neither human action and agency nor identity and subjectivity can be properly understood independent of nonhuman entities and their agencies. To frame this in phenomenological language, we "go back to the things themselves" through our lived experiences of them, and we find the "inner core of subjectivity" (Coole 2001) there, in the very material, intercorporeal interworld of more-than-human interconnections.

This is also how Merleau-Ponty's broader contribution to phenomenology can be interpreted. He developed a body-centered methodology precisely to "go back to the things themselves." While both things and experiences are seen as ideal forms and inner mental states in the classical phenomenology of Husserl (Carman 2012), Merleau-Ponty materialized the phenomenological conception by tying perception and experience firmly to the lived body, by placing embodiment at the heart of phenomenological inquiry (Landes 2012). In doing so, he turned to sensation, not as a purely subjective experience, but as "the living relation of the one who perceives with both his body and his world" (Merleau-Ponty 2012, 216).

Tim Ingold argues, with reference to James Gibson, for an ecological approach to perception, which does not treat the mind as "a data processing device" that organizes and interprets "the raw material of experience consisting of sensations," but as "immanent in the network of sensory pathways that are set up by virtue of the perceiver's immersion in his or her environment" (Ingold 2000, 2–3). Merleau-Ponty's later work is akin to such an ecological approach to perception, as "he no longer takes consciousness as starting point. . . . [H]e begins with nature to show the identity in it of being and being-perceived" (Barbaras 2001, 37). Perception, thus, is not an act of an isolated consciousness, of a perceiving subject; it is the bond between bodily beings, the positions of which within the bond of perception are reversible, and within which embodied subjectivities and agential capacities arise. Focusing on this bond, on our bodily sensory and affective connection with the world, provides us with a firm methodological tool with which to explore human agency within "an ecology of human and nonhuman elements" (Bennett 2010, 103).

This book employs such a phenomenological perspective, not because I, as its writer, happened to be a phenomenologist, but because I followed the ethnographic data, which demanded a conceptual vocabulary to discuss "how agentic properties emerge and endure within corporeal experience" (Coole 2005, 131). One of the main arguments of the book, in a nutshell, is that East Black Sea women's radical political agency against HEPPs is conditioned by the intimate corporeal connections they have with river waters, and by bodily memories of such connections that are anchored in place. Their statements emphasize, over and over again, an interconnectedness, "an indivision," with the rivers, which makes the cause of the anti-HEPP struggle vital and urgent for them.[3] The state of interconnectedness is shaped by a series of sensory and affective encounters with the nonhuman entities and environments—in our case, with the river waters. Women's resistance to HEPPs could thus be thought of as a "carnal resistance" that "initiates transformative acts" (Coole 2005, 131), with the bodily senses, affects, and emotions acting as the media of political agency. In other words, the resistance is carnal and embodied not only as it is performatively enacted by the protesting bodies of women, but also because it is engendered through bodily processes of sense, affect, and emotion within a more-than-human lifeworld.

Ben Anderson and Paul Harrison define nonrepresentational theory as an approach that finds the roots of action less in willpower and cognitive deliberation and more in "embodied and environmental affordances, dispositions and habits" (Anderson and Harrison 2010, 7). The case of the anti-HEPP movements in the East Black Sea Region demonstrates that "embodied and environmental affordances, dispositions and habits," as well as corporeal experiences, are at the root of *political* action as well. Such an understanding—one that goes beyond a limited notion of the political as a field of abstract ideas and interests,[4] that grounds political agency in the realm of the intercorporeal (Fielding 2017), in which everyday practices, experiences, and encounters involve both human and nonhuman elements—is crucial if we are to accurately and adequately assess the political and environmental struggles of our age.

This book contributes to the effort of thinking politics from the perspective of the body by demonstrating the centrality of corporeal experience, sense, and affect to the way anti-HEPP opposition is framed and enacted. Such an approach, I firmly believe, opens new pathways for studying women's political subjectivity within and beyond environmental struggles. It maintains, among other things, that the embodied subjectivity is formed out of intercorporeal interaction not only with other humans, but also with nonhuman bodies and things. In other words, the world of encounters in which our subjectivities emerge is not limited to human beings: the "intersubjectivity that is at the root of being in the world includes many different kinds of subjects, of which humans are but one subset" (Whitmore 2018, 23). The subject, then, is not abstract, self-centered, unitary, and autonomous, but embodied, sentient, interdependent, and transversal; it is a subject that "functions in a nature–culture continuum" (Braidotti 2013, 61).

Not only do embodied subjectivities and agentic capacities function in this nature-culture continuum, but the everyday lifeworlds from which they emerge do as well. Environmental entities—the rivers in our case—are much more than "natural resources"; they are integral and essential to the place-based lifeworld. Kay Milton (2002) writes about the powerful alliance between science and economics that leads us to treat nature and natural things as resources. Grassroots struggles for environmental commons, at least in cases like those in the East Black Sea Region, contradict this instrumental representation of nonhuman nature. In this sense, these struggles cannot simply be framed as "resource conflicts" or "resource struggles." They are, instead, struggles against such a framing, against the reduction of nature and the natural entities that constitute not just their *umwelt* (environment, surroundings), but also their *lebenswelt* (lifeworld), to mere resources that can be owned and traded.[5] To put it differently, these struggles point to the ways in which the ecological spills over into what we call the social, as captured with the notion of *socio-ecological justice* that I have developed in chapter 5.

As the ecological spills over into the social and as nonhuman entities are an essential part of social interactions in place, *eco-sociality* (Whitmore 2018) emerges as central to the framings, motivations, and justice claims of local communities fighting for environmental commons. I have demonstrated throughout the book how living in the riverside villages is *living with* a river, as people work, unwind, celebrate, socialize, sing, cry, fall in love, always in constant bodily interaction with river waters. Such constant bodily interaction engenders a close and intimate relationship, a feeling of connectivity, of *Einfühlung*, which is manifested in the narratives of rivers and river waters as vital and essential. The river in these narratives is not a mere resource, but a companion: the sight, the sound, and the touch of life itself that affects bodies, relationships, places, and memories. To put it differently, the agency of nonhuman entities, here in the form of rivers, is not limited to shaping the physical and economic space; it is constitutive of sociality, history, and identity.

It is now common wisdom in critical geography, science and technology studies, and environmental humanities not to treat the social and the ecological (or society and nature) as two separate realms of human and nonhuman relations (see, e.g., Haraway 1991; Latour 1993; Swyngedouw 1999; Whatmore 2002; Descola 2013). The notion of socio-ecological justice transmits this wisdom to the field of environmental struggles and environmental justice by framing human-nonhuman relationality as a matter of justice. In doing so, socio-ecological justice speaks for this experience of *living with* nonhuman entities that is reflected in the justice claims of grassroots movements, relating these claims to the "already existing" (Thomas 2015) relational ontologies and ethics of *coexistence*. This ethics and experience of coexistence goes against "scientific" descriptions of environmental entities as "resources" that reflect an instrumental and hierarchical relation between human and nonhuman beings—what Lorraine Code (2006) has called "epistemologies of mastery."

Starting with the lived body and arriving at socio-ecological justice, this book bridges feminist phenomenology and ecological thought in light of the relational ontology of Merleau-Ponty's later work. While feminist phenomenology fleshes out and situates the lived body within a social, sexual, historical, and cultural field of power and difference, ecological thinking extends the situation further to underscore the ecological embeddedness of our social existence. We are thus not only socially, not even only socio-spatially, but socio-ecologically situated as lived bodies. The socio-ecological relations we are situated in are central to the processes of subjectivization. When the subject is socially and ontologically defined in continuity with the world through its corporeal connection with it, subjectivity can only be envisioned as ecological, relational, and transversal—contingently and temporally emerging as a moment of singularity within the materiality of the world, through perceptive and affective experiences.

The question of ecology, hence, is not a matter of conserving a nature "out there." Ecological thinking is mainly about thinking transversally, which leads to what Guattari calls an *ecosophy*, that is, an "ethico-political articulation" between "three ecological registers" of nonhuman entities, social relations, and human subjectivity (Guattari 2000, 28). This book aims at making a modest contribution to such an articulation, to an ecological conception of lifeworld, sociality, and subjectivity, by incorporating the empirical potential of an ethnographic case study and the conceptual promises of a vigorous transdisciplinary approach.

Appendix

Turkey comprises seven geographical regions, twenty-one subregions (East Black Sea, West Black Sea, and Central Black Sea Regions are subregions of the Black Sea Region, whereas the Mediterranean and East Anatolia are two of the seven main regions), eighty-one provinces, 973 districts, and more than 30,000 villages. The table below aims to provide the reader with detailed information about the geographical location of almost every village, valley, and district that is mentioned in the book. As the table demonstrates, most of the villages mentioned are located in valleys, valleys are located within the borders of districts, districts constitute provinces, and every province is a part of a region.

Table 1   Geographical Locations of the Villages, Valleys, and Districts
Mentioned in the Book

Villages	Valleys	Districts	Provinces	Regions
Arılı	Arılı	Fındıklı	Rize	East Black Sea
Gürsu	"	"	"	"
Yaylacılar	"	"	"	"
Meyvalı	"	"	"	"
Aslandere	Çağlayan	"	"	"
Rüzgarlı	İkizdere	İkizdere	"	"
	Senoz	Çayeli	"	"
		Çamlıhemşin	"	"
Konaklı		Arhavi	Artvin	"
Ulukent	Pilarget	"	"	"
Balıklı	"	"	"	"
	Kamilet	"	"	"
		Ardanuç	"	"
Camili	Macahel	Borçka	"	"
Meydancık	Papart	Şavşat	"	"
Bazgiret	"	"	"	"
		Hopa	"	"
	Solaklı	Of	Trabzon	"
		Tonya	"	"
Engiz		Fatsa	Ordu	"
Çiftlik		Korgan	"	"
Düzköy		Çanakçı	Giresun	"
	Loç	Cide	Kastamonu	West Black Sea
	Yeşilırmak		Amasya	Central Black Sea
Pınar	Yuvarlakçay	Köyceğiz	Muğla	Mediterranean
Beyobası	"	"	"	"
Ahmetler	Ahmetler	Manavgat	Antalya	"
Karacaören	Alakır	Kumluca	"	"
Boğazpınar	Karasu	Tarsus	Mersin	"
Çamlıkaya	Aksu	İspir	Erzurum	East Anatolia
Bağbaşı	Ödük	Tortum	"	"

Notes

INTRODUCTION

Throughout the book, I indicate the regions in which the villages and valleys referred to are located, to give the reader a general idea of their geographical position. I use the following abbreviations: EBR for the East Black Sea Region; WBR for the West Black Sea Region; CBR for the Central Black Sea Region; MR for the Mediterranean Region; and EAR for the East Anatolia Region. See the appendix for more precise information regarding the districts and provinces in which they are located.

1. GAP is a highly ambitious, large-scale development project involving twenty-two dams and nineteen hydroelectric plants throughout southeastern Turkey. It targets the presumably "backward" Kurdish region and is based on a rather colonial approach that equates hydroelectricity and irrigation with civilization, modernization, and development. Not surprisingly, it has led to controversies and conflicts in both the national and transnational political arenas (Çakıroğlu and Eder 2001; Harris 2002; Eberlein et al. 2010; Hommes, Boelens, and Maat 2016; Bilgen 2018).

2. Run-of-the-river HEPPs use micro turbine generators to capture the kinetic energy of the natural flow of water running down steep slopes. Typically, water from upstream is diverted to electricity-generating turbines by a weir or a pipeline, and then released back into the river's downstream flow.

3. John Lovering and Hade Türkmen use the term *bulldozer neoliberalism* in the context of urban renewal in Istanbul, but the bulldozer is an apt symbol for the rule of the AKP far beyond this specific example, as the party has based its agenda on a synthesis of authoritarian neoliberal rule and extractivism (Adaman, Arsel, and Akbulut 2019).

4. Construction of private and small-scale HEPPs became possible in Turkey only in the 2000s after an open energy market was established, a condition set by the IMF and the World Bank to release credit after the economic crisis of 1998–99 (Atiyas, Çetin, and Gülen 2012).

5. It now seems impossible for the AKP government to reach this number by 2023, as many projects have been thwarted by local communities, with the support of court decisions. In some cases, the companies themselves canceled projects, due to lack of feasibility, changing economic circumstances, and other case-specific factors. According to the DSI (State Hydraulic Works) 2021 Activity Report, 743 HEPPs were in use at the end of 2021, 675 of them being private and small-scale. While 26 HEPPs are under construction, 469 more are in the project phase, which would bring the total number to 1,238 if all the planned projects were completed, which is highly unlikely. See https://cdniys.tarimorman.gov.tr/api/File/GetFile/425/Sayfa/759/1107/DosyaGaleri/2021_yili_faaliyet_raporu.pdf, last accessed on April 10, 2022.

6. This extensive development boom targets, more or less, every single stream in the country. The government's plan to exploit 10,000 km of river systems for HEPP and hydro dam development leaves no room for natural ecosystems to function. There are 185 of 305 key biodiversity areas in Anatolia under the threat of HEPP development, which endangers natural habitats and could lead to the extinction of many species. It has gone so far that anti-HEPP activists have started to use the term *genocide of the rivers* to capture the extent to which the country's natural river ecosystems, natural habitats, fish and wildlife passages, biodiversity, and forestry have been disturbed (Turkish Water Assembly 2011; TMMOB 2011; EMO 2011; Başkaya, Baskaya, and Sari 2011; Şekercioğlu et al. 2011; Kurdoğlu 2016).

7. On water commodification and dispossession, see Ahlers 2010; Ahlers and Zwarteveen 2009; Bakker 2003.

8. Local communities (predominantly rural/peasant, Indigenous, Black, and minority) all around the globe are resisting various forms of land and water grabbing and are fighting against extractivist projects (mining, deforestation, extraction of oil from tar sands, fracking, etc.) and contamination of water and land by toxic waste.

9. I conducted semistructured, in-depth interviews (see Blee and Taylor 2002) with the members and representatives of local associations, as well as regional and national platforms such as DEKAP (*Derelerin Kardeşliği Platformu*—Brotherhood of the Rivers Platform), KIP (*Karadeniz İsyandadır*—

Black Sea in Resurrection Platform), and MEH (*Mezopotamya Ekoloji Hareketi*—Mesopotamian Ecological Movement).

10. This includes content analysis of visual material (videos and photographs) and written texts produced by protesting communities and anti-HEPP associations and platforms (usually made available through social media), hundreds of news stories published in national media outlets, independent archives of local environmental organizations such as the *Ekoloji Almanağı* (English title: *Ecology Almanac*) (Aksu and Korkut 2017), and documentaries by independent film makers.

11. Needless to say, there cannot be a strict differentiation here, as some villagers act as local activists, and some local activists have roots in the villages.

12. At the time of my field research, Kamile Kaya was the only woman on the board of *Derelerin Kardeşliği Platformu* (Brotherhood of the Rivers Platform), which is the primary national/translocal network of the anti-HEPP movement with connections across the country. In all the places I visited, I met only two women, Seniye and Esin, who worked actively on the boards of their local anti-HEPP organizations—in Fındıklı (in the EBR) and Alakır (in the MR).

13. Georgina Drew (2017) tells a funny story of a male neighbor who visited her to share his knowledge about the Ganga (Ganges River) and its religious importance in relation to the antihydropower struggles in the region. "I am a priest, so I know everything there is to know about the subject," he said, and "immediately launched into a discussion on the Ganga's role in various historical Hindu texts and scriptures." When Drew told him that she was "increasingly interested in how the women of the region understood Ganga's importance, he scoffed audibly. . . . 'Women,' he stated, 'do not know anything about the Ganga.'" This anecdote demonstrates the prevalence of men's insistence on imposing their views, valuing knowledge that references a higher order such as religion, science, or politics (domains that tend to be dominated by men) while belittling the practical, experiential knowledge women acquire through practices of everyday living and dwelling.

14. What made me appear young—"like a student"—was not necessarily my age or physical features, but my style of clothing, consisting of jeans or trekking trousers, simple T-shirts, and a rain coat, which, in their minds, would be inappropriate for a university lecturer.

15. On the War of Independence (1919–1923) and the modern history of Turkey, see Zürcher 1993.

16. The exact word they use in Turkish is *dere*, which could be translated more accurately as "stream" than as "river." The streams they are talking about tend not to be very long or deep (the Arılı, for example, is 31.5 km long). Run-of-the-river HEPPs are usually built on midsize streams that are steep and fast flowing, and which do not carry large volumes of water. I choose to use the term *river* throughout the book, however, for the sake of consistency, as I refer to many streams and rivers of different sizes. By calling them rivers, I draw on the definition of a river as a natural flow of running water that follows "a well-defined,

permanent path, usually within a valley," to differentiate it from streams, brooks, and creeks, which may have more temporary courses and which tend to disappear in certain seasons. See https://www.encyclopedia.com/science /encyclopedias-almanacs-transcripts-and-maps/river-and-stream, last accessed on March 18, 2021.

17. This does not mean, of course, that the body is constitutive of the world as such. It is constitutive of *our* world, of the world as we experience it.

18. Besides Elizabeth Grosz (1987, 1994), these pioneers include Iris Marion Young (1980), Moira Gatens (1988, 1996), Rosi Braidotti (1991, 1994), Vicki Kirby (1997), Gail Weiss (1999), Abigail Bray and Claire Colebrook (1998), and Toril Moi (2001).

19. It also does not, in principle, exclude the experiences and narratives of people who do not identify themselves within the binary heterosexual system of sexual orientation. I did not, however, encounter any openly LGBTQIA+ people during my field research. In this sense, this study only de facto excludes their experiences and views due to lack of data.

20. While phenomenology, especially in its critical and feminist readings, has proven useful in developing "a robust account of embodied lived experience" (Fielding 2017, ix; see also Olkowski and Weiss 2006; Guenther 2013; Fielding and Olkowski 2017; Salamon 2018; Simonsen and Koefoed 2020), new materialist and eco- and postphenomenological approaches (e.g., Brown and Toadvine 2003; Coole and Frost 2010; Ash and Simpson 2016) maintain the relational, intercorporeal, and more-than-human character of the intersubjective interworld within which lived experience emerges.

21. The East Black Sea Region is located in the northeast of Turkey, bordering Georgia. Cities and towns are squeezed into a narrow coastal strip between the Black Sea and the Kaçkar Mountains that run parallel to the Black Sea coast. (The Kaçkar Mountains are also known as Pontic or North Anatolian Mountains, but in this book I will use the name the locals use, Kaçkar). The rural population lives in villages located in the densely forested, steep valleys of the region. The relatively small but immensely impressive, fast-flowing, and cascading rivers of the region typically flow from the mountains to the sea along those valleys. The geography and the spatial organization of the region will be discussed in detail later in the book.

22. Between 2008 and 2018, 203 HEPPs were built in the Black Sea Region, concentrated in the eastern parts of the region, while 20 were under construction and 123 more were in the project phase as of 2018. See https://www .hurriyet.com.tr/gundem/karadeniz-hes-denizi-oldu-10-yilda-203-hes-yapildi-143u-yolda-40711636, last accessed on March 18, 2021. By August 2021, the total number of active HEPPs in the region was 246; see https://m.bianet.org /bianet/yasam/248676-karadeniz-de-246-aktif-hes-var, last accessed on April 10, 2022.

23. This is not to suggest that all the communities in the region where HEPP projects have been constructed or proposed are committed to blocking these projects. While the region is home to symbolic cases of the anti-HEPP struggle, such as Fındıklı, where the residents are united against the HEPP projects, there are also localities in which no opposition has emerged, and others in which the local community has been divided into two camps. See the next chapter for a detailed discussion.

24. Merleau-Ponty's concept of the flesh can roughly be defined as the vast network of Being, the common web of bodies, things, organisms, ecosystems, etc., shared by persons and things, in which tactile and sensory relations occur (Merleau-Ponty 1968, 2003; see also Grosz 1994; Butler 2005; Olkowski 2006; Coole 2007). A more detailed discussion of the concept can be found in chapter 3.

25. Even though environmental problems and conflicts are becoming increasingly pressing in the region, the whole issue is still discussed, especially in the case of water, within the grand framework of scarcity, security, and policy, with little emphasis on the struggles on the ground (see, e.g., Ward and Ruckstuhl 2017; King 2020). This tendency is slowly changing, though, and environmental movements have been getting more attention (Verhoeven 2018; İnal and Turhan 2020).

26. Greetings to my friends and colleagues who have been problematizing this in the Theorizing through the Global South reading/working group, based at the Institute for Cultural Inquiry (ICI) in Berlin.

CHAPTER 1

I have translated *Derelerin Kardeşliği Platformu* (DEKAP) as "Sisterhood of the Rivers Platform" in previous publications, as the word *kardeş* ("sibling") does not designate a particular gender (Turkish is not a heavily gendered language in general). A literal translation would be "Fraternity" or "Siblinghood of the Rivers Platform." But as it is generally called the Brotherhood of the Rivers Platform in English, as in the quoted article, I adopted this use to prevent confusion.

1. Mustafa Kemal Atatürk's tenth anniversary speech is available in English at http://www.columbia.edu/~sss31/Turkiye/ata/onuncuyil.html, last accessed on August 10, 2022. Due to space limitations, it is unfortunately not possible to provide a detailed historical context regarding the establishment of the Turkish Republic. For accounts of Turkish modernization, Mustafa Kemal Atatürk, and the establishment of the republic, see Berkes (1964); Zürcher (1993); Bozdoğan and Kasaba (1997); and Kerslake, Öktem, and Robins (2010).

2. For a detailed discussion of the Justice and Development Party, see Baykan (2020).

3. For the Turkish government's letter of intent promising to privatize the energy sector in the context of its request for financial support from the IMF, see

https://www.imf.org/external/np/loi/1999/120999.htm; see also Atiyas, Çetin, and Gülen (2012).

4. For a detailed discussion of the AKP's "socio-environmentally destructive megaprojects, transportation and infrastructural investments," see İnal and Turhan (2020).

5. For instance, private companies pay only 1 percent of the license fee at the time of the agreement and can pay the remaining amount over a period of eight years. Moreover, the government guaranteed it would buy the full amount of electricity produced by small-scale private HEPPs for US$7.30 per kilowatt-hour for at least ten years. In addition to such incentives, small-scale and private HEPP projects are financially supported by a range of domestic and international renewable energy and climate change funds, such as the Clean Technology Fund of the World Bank, as well as credits and grants from various financial institutions (World Bank 2011; Turkish Water Assembly 2011; Işlar 2012).

6. For studies that frame run-of-the-river HEPPs as the key to clean and renewable energy development in Turkey, see Yuksek et al. (2006); Çapik, Yılmaz, and Çavuşoğlu (2012); Yüksel (2013).

7. Quoted by CNN Türk, August 11, 2010, translation by the author.

8. In 2018, the Ministry of Forestry and Water Affairs and the Ministry of Agriculture were united as the Ministry of Agriculture and Forestry.

9. Eroğlu, quoted in the daily newspapers *Birgün* (October 12, 2014) and *Milliyet* (May 12, 2015).

10. Eroğlu, quoted by Ihlas News Agency (November 7, 2010).

11. Erdoğan Bayraktar, who served as minister of Environment and Urban Planning from 2011 to 2013, and Taner Yıldız, who was the minister of Energy and Natural Resources from 2009 to 2015, both publicly acknowledged the negative effects of small-scale HEPPs on river ecosystems and natural habitats.

12. Quoted in the daily newspaper *Radikal* (August 22, 2008).

13. Erdoğan, quoted by CNN Türk (August 11, 2010).

14. *Sudaki Suretler*, directed by Erkal Tülek, can be accessed in Turkish at https://archive.org/details/SudakiSuretler, last accessed on March 16, 2021. Watching these documentaries might help the reader envision the everyday practices of people as well as the geographies of villages in which struggles against hydropower plants take place. For this particular documentary, English subtitles are, unfortunately, not currently available.

15. For ongoing campaigns, struggles, and discussions regarding the right to water, see Right2Water Campaign (http://www.right2water.eu/about-our-campaign).

16. In practice, however, *everyone* mostly indicates a form of *we;* see Ostrom 1990.

17. All quotations are taken from interviews and recorded conversations conducted by the author, unless stated otherwise. Only first names of interviewees

are used in the text; in some cases, even first names are pseudonyms. Surnames are only used when they are already public, such as in interviews published in newspapers or books or appearing in video reports. All translations from Turkish are by the author.

18. See Harvey's *The New Imperialism* (2003, 145) for the definition and a detailed discussion of "accumulation by dispossession."

19. Mezzadra and Neilson (2019) define this current phase of capitalism as "extractive," referring to processes of wealth extraction both from the earth— "literal removal of raw materials and life forms"—and from patterns of human cooperation and sociality.

20. Turkey is not a part of the regulatory carbon credit market. Carbon-offsetting projects in Turkey, however, are benefitting from voluntary emissions trading. Turkey ranked as the world's sixth-largest carbon offset supplier in 2016, with 218 registered projects, most of which are hydropower plants (Turhan and Gündoğan 2019).

21. Erdoğan supporters call him *Reis* ("the Chief"), a nickname he apparently embraces.

22. This documentary is available without English subtitles at https://www .youtube.com/watch?v = oHdBOAhiwww, last accessed on March 18, 2021.

23. These benefits range from open bribes to job offers, offers to lease any machinery (such as trucks) the village people might possess, and promises to build a new school or a mosque in the village or to repair roads and infrastructure.

24. Environmental assessment is defined by the European Commission as "a procedure that ensures that the environmental implications of decisions are taken into account before the decisions are made." The EIA "can be undertaken on the basis of Directive 2011/92/EU (known as 'Environmental Impact Assessment'—EIA Directive) for individual projects, such as a dam, motorway, airport or factory or for public plans or programmes on the basis of Directive 2001/42 /EC (known as 'Strategic Environmental Assessment'—SEA Directive)." See https://ec.europa.eu/environment/eia/index_en.htm, last accessed on May 23, 2022.

25. The word she uses for "consent" is not the usual *rıza* but *razılık*, which is specifically used within Alevi theology as one of the requirements of *Cem* (religious/communal ceremonies of coming together and worshiping), which means that every member of the *Cem* community should be content with one another and with the composition of the whole group. *Sudaki Suretler,* by Erkal Tülek, can be accessed and watched at https://archive.org/details/SudakiSuretler. English subtitles are unfortunately not currently available.

26. It seems that this struggle has been lost, as construction of the Ilısu Dam has continued despite all the opposition, and filling of the reservoir started in the last few months of 2019. The first months of 2020 bore witness to very sad

moments, as a town with 12,000 years of history slowly disappeared under the rising waters. See the *ARD Weltspiegel* reportage at https://www.youtube.com /watch?v = cSuxCtak-C4 (in German), last accessed on March 16, 2021.

27. See https://www.cumhuriyet.com.tr/haber/ocalandan-akpye-uyari-sabir-tasi-catladi-101695, last accessed on March 16, 2021.

28. This categorization does not explain all the existing cases, though; certain parts of the East Anatolia and East Black Sea Regions (the inland area of the East Black Sea Region and its neighboring parts in East Anatolia, such as Tortum and İspir in Erzurum) share similar characteristics with the Mediterranean Region in terms of the use of river waters for irrigation. Also, even though the majority of East and Southeast Anatolia's population is Kurdish, there are also Turkish communities there that resist HEPPs, as in Tortum and İspir in Erzurum, with different motivations from the Kurdish people who resist HEPPs in Dersim and Hasankeyf.

29. One should acknowledge here the recent attempts to recognize and discuss the historical roots, cultures, and identities of Laz and Hemshin peoples (Simonian 2007; Biryol 2012, 2014). The *Biryaşam* ("Onelife") association in Hopa (Artvin Province) is a valuable local example, as it brings issues of ecology and identity together and openly discusses the Armenian roots of Hemshin people. These attempts do not have enough public influence to challenge the strong association of Laz and Hemshin peoples with Turkish political identity, though.

30. One exception is the Romeyka- (Pontic Greek) speaking people of Ogene in the Solaklı valley of Trabzon, for whom, argues Oğuz (2016), Derebaşı (the name given to the place where the headwaters of the Solaklı River are located) is a sacred place, a place of wailing for the dead.

31. As might be expected, the way river water is used depends on geographical location. In the terrestrial inland part of the EBR, which is located toward the East Anatolia Region, at the southern edge of the East Black Sea mountain range, rainfall is much less than in the towns located in the coastal area on the northern side of the East Black Sea mountain range, facing the Black Sea. Thus, for example, in one of the towns I visited during my field trip in the region— Ardanuç and its neighboring villages Tosunlu and Bulanık—river waters are often used for agriculture and animal husbandry, unlike in many other places in the East Black Sea Region. The most well-known cases of the anti-HEPP struggle, though, such as Fındıklı, are located in the coastal part, where river waters are not utilized for immediate economic purposes.

32. The documentary film *İşte Boyle* (*Damn the Dams*), made by Özlem Sarıyıldız in 2012, is not fully accessible online. A long trailer with English subtitles can be watched through Vimeo: https://vimeo.com/37968420.

33. I make this claim on the basis of the interviews I conducted with five members of the CEHAV network—Bora Sarıca, Berna Babaoğlu, Cömert Uygar Erdem, Barış Yıldırım, and Fevzi Özlüer—from 2013 to 2015 in Ankara, Istan-

bul, Dersim, and Fethiye. This view also comes across strongly in the published interviews of Yakup Okumuşoğlu and Fevzi Özlüer, conducted by Sinan Erensü (Erensü 2016a, 2016b).

34. The passing of the Draft Law, which manifests the AKP government's will to eliminate the environmental protections that slow down its extractivist strategy of growth (see, e.g., WWF 2013), was impeded by environmentalist NGOs and local environmental movements for more than a decade.

35. This is, of course, not unique to Turkey. From Nigeria to the Philippines to Guatemala, in many different countries in the so-called Global South, police and military act as the de facto armed forces of private companies (Nixon 2011; Henighan and Johnson 2018).

36. The PKK (Kurdistan Workers' Party) has been involved in an armed conflict with the Turkish state for more than three decades and is normally a figure of hatred and fear in all parts of Turkey except the Kurdish region.

37. See Yeğen (2007) for the position of Turkish nationalists regarding the Kurdish question. See also Bora (2003) and Nefes (2018) for a detailed discussion of Turkish nationalism and nationalist discourses.

CHAPTER 2

1. *Yayla* is the Turkish word for summer settlements and pastures on the high plateaus of the Black Sea Mountains. Traditionally, East Black Sea villagers move, mostly on foot, to those higher settlements in the summer with their families and animals.

2. Military service is compulsory for every healthy man in Turkey.

3. The interview is available at https://www.cumhuriyet.com.tr/haber /sari-yazmali-halime-cakmak-hava-su-toprak-otesi-var-mi-493114, last accessed on August 25, 2022.

4. See https://www.hurriyet.com.tr/gundem/tortumda-hes-protestosu-18831594, last accessed on June 16, 2022.

5. On the Turkish War of Independence, see Erickson (2021).

6. Women's narratives are presented and discussed in detail in the coming chapters.

7. Despite stressing the centrality of everyday bodily interactions to women's connection to the river Ganga and their opposition to HEPP projects, Drew does not translate these ethnographic observations into a conceptual analysis in terms of developing a body-centered framework to study women's agency and activism.

8. As clarified later in the chapter, this is not an essential difference in either Garhwal in India or the East Black Sea Region of Turkey. Rather, the gendered division of labor; access to scientific, political, and public discourse and education;

degree of contact with state apparatuses and authorities; and patterns of interaction with river waters are reflected in the discourses and narratives of men and women.

9. *Le Reprise Du Travail Aux Usines Wonder* ("Resumption of Work at the Wonder Factory"), made by Pierre Bonneau, Liane Estiez-Willemont, and Jacques Willemont, is accessible at https://vimeo.com/276078088 with English subtitles, last accessed on March 16, 2021.

10. The moniker "Crazy HEPP" is most likely a reference to the Istanbul Canal, Erdoğan's so-called crazy project, which would also have substantial socio-ecological impacts similar to the Kavak HEPP.

11. It is a common strategy of HEPP companies to lease construction machinery and trucks from locals at higher than market prices as a way of gaining their consent.

12. The women saying that they did not want the money were not at all rich. The house we were sitting in front of did not have paint or siding; one could see the bare bricks. But they did not live in absolute poverty, either, in the sense that they could afford decent food, and most of them could support their children throughout their studies, albeit with difficulty.

13. *Selling the river* is a commonly used term to refer to the transfer of river waters to private companies.

14. Fındıklı (center) is a seaside town, but the sea is still a car journey away from people who live up in the villages. In addition, Black Sea coastline was radically altered with the construction of the Coastal Highway in the first half of the 2000s, which disrupted the town centers' connection and access to the sea. If we add the rapid pollution of the Black Sea to the picture, it becomes clear that the sea has become a lost cause for the peoples of the region.

15. Ecofeminism is a school of feminist thought that establishes a close connection between the domination of women and the domination of nature by patriarchal and colonial Western culture and/or capitalism (see, e.g., Merchant 1980; Gaard 1993; Plumwood 1993; Mies and Shiva 1993; Sturgeon 1997). While early variants of ecofeminism imply an essential, elemental, or biological affinity between women and nature, certain (later) versions of ecofeminism stress the "socially constructed" character of the women-nature connection (see, e.g., Eaton and Lorentzen 2003; Mellor 2003; Gaard 2017). For recent interpretations and ecofeminist engagements with posthumanist and new materialist perspectives, see Philips and Rumens (2015); Merrick (2017); Thomas and MacGregor (2017); Vakoch and Mickey (2017); and Ya-Chu Yang (2017).

16. Feminist political ecology (FPE) is probably the most significant and influential of the approaches mentioned above. FPE was born out of an attempt to unite political ecology's focus on control of, and right of access to, "natural resources" with a feminist perspective to maintain gender as a "critical variable

in shaping resource access and control" (Rocheleau, Thomas-Slayter, and Wangari 1996; see also Nightingale 2006; Harcourt and Nelson 2015).

17. The labor force participation rate in nonagricultural sectors among women in the East Black Sea Region is 17 percent. The majority of women who work in the agricultural sector are unpaid, as they work on their own household's land (Gazioğlu 2014; Yaman 2019).

18. When the women of the household are not able to provide sufficient labor for harvesting, seasonal workers are hired, mostly from neighboring Georgia, and, interestingly, mostly men.

19. Karaçimen and Değirmenci (2019) state that men tend to work with women in the tea fields if tea farming is the only source of income for the family, which is now a rare case in the region. They also found, through their fieldwork in different towns and villages of Rize Province, that the thoroughly gendered division of labor in tea farming tends to be reproduced by new generations even when they are university students or graduates—that is, girls work in the fields with their mothers during their summer holidays, while boys deal with more logistical issues such as transportation and transactions. Their findings are in line with my own observations.

20. Only university graduates are eligible for the shorter-term military service (six months).

21. See https://www.youtube.com/watch?v = Qr79ZtFQLYk, last accessed on June 6, 2022.

22. Throughout my fieldwork it was mainly men who questioned me in different ways, wanting to ensure that I was not secretly working for the HEPP companies or reporting my informants to the state. Their cautious and suspicious attitude, as uncomfortable as it is for me, is not entirely unfounded. It is common knowledge that private companies send their representatives to the villages, sometimes hiding their identities. And anyone can report anyone in Turkey for "antigovernment activities," which is typical of authoritarian regimes and can result in lawsuits and even prison sentences.

23. This feminist argument is also in line with the broader field of political ecology, which is dominated by a strong assumption that the environmental struggles of local (rural, native/Indigenous) communities are driven by immediate dependence of these communities on "natural resources" (see, e.g., Martinez-Alier's [2002] *The Environmentalism of the Poor*). Recently, a second motivation has also been acknowledged, that of protecting cultural identities rooted in place-bound cosmologies and beliefs, especially in the case of native/Indigenous and/or minority communities (see, e.g., Martinez-Alier et al. 2016).

24. An exception is the terrestrial inland part of the region, which is located toward the East Anatolia Region, at the southern edge of the East Black Sea mountain range, where the rainfall is much lower than in the towns located in

the coastal part. In those inland parts, river waters might be used for irrigated agriculture and animal husbandry. That does not challenge my argument here, however, as the majority of local communities that oppose HEPPs are located in coastal parts where livelihoods do not depend on river waters.

25. Beyond phenomenological geography, this basic theme of the "making of the subject" within and through relations with immediate environments, i.e., the objects, things, and organisms around which practices of work are structured, is central to many theories of the social, from Marx to Simmel.

26. Drawing on Georgina Drew's work, one could make the same argument regarding the religious value and sacredness of river waters. Drew demonstrates that even though Hindu religion, traditions, and mythology provide the context for the relationship between human communities, especially women, and the Ganges River as a goddess, they do not determine "the nature of human connection and care for the river." As Drew (2017, 16–17) explains, "Ganga's framing in Hindu mythology might alert people to the goddess's importance, but it is the cumulative effect of their everyday interactions and connections with the divine entity that can solidify means and manners in which the river obtains its deepest significances."

27. It does not follow, of course, that all men and women are the same in terms of how (and how much) they are connected to their immediate surroundings and to river waters. I have no intention of constructing uniform binaries here (such as men and women); rather, I aim to reveal certain systematic and gendered patterns that organize our relations with our human and nonhuman environments on the basis of our (assumed) gender identities.

28. An important exception to this trend is, of course, Stacy Alaimo, who developed the widely cited notion of transcorporeality, which refers to the idea that "the human is always intermeshed with the more-than-human world" and underlines "the extent to which the substance of the human is ultimately inseparable from 'the environment'" (Alaimo 2010, 2).

29. The most influential term of feminist theory, *gender,* is employed to avoid the essentializing associations between nature, female biology, and women's inferiority. Feminists argue that there is nothing "natural" to the subordination of women; it is socially constructed through patriarchal institutional structures and relations of power. The sex/gender distinction has been widely adopted within and beyond feminist circles in both academic and public spheres. It has also been criticized, though, by many feminist scholars from very different perspectives, for reproducing the binaries that feminism has tried so hard to destabilize (see, e.g., Gatens 1988, 1996; Butler 1990; Moi 2001; Young 2005; Colebrook 2017).

30. Feminist phenomenologists have operationalized Merleau-Ponty's concept of the "lived body" to do what he left undone: to articulate the body as differentiated, "to describe embodied being-in-the-world through modalities of

sexual and gender difference" (Young 2005, 7; see also Olkowski and Weiss 2006).

31. The concept of being-in-the-world comes from Heidegger (*In-der-Welt-sein*) and denotes the embeddedness of *Dasein* ("being"; "existence") in an everyday, material world.

32. My interpretations of Ana Mendieta's earth-body sculptures are based on an exhibition titled *Covered in Time and History: The Films of Ana Mendieta*, exhibited at the Gropius Bau in Berlin (April 20–July 22, 2018).

33. See Kandiyoti (1988) for the concept of "bargaining with patriarchy."

34. *Teyze* ("aunt") is a respectful way of addressing an older woman with whom one has developed a close relationship.

CHAPTER 3

1. See the section "Corporeal Feminism, Merleau-Ponty, and the Lived Body as a Missing Link to Relate Gender and Environment" in the previous chapter for a discussion of the relationship between the anonymous body and the lived body.

2. Bromber, de la Croix, and Lange (2014) define rivers as "local at all points" with reference to Bruno Latour, who used the same expression for railroad tracks.

3. See chapter 2 for a detailed discussion of the gendered division of labor in the EBR.

4. Throughout the book, when I cite Merleau-Ponty 2012, I am referring to the new English translation of *Phenomenology of Perception* by Donald A. Landes (Routledge 2012). The French original was published in 1945.

5. The reportage can be viewed at https://www.youtube.com/watch?v = 8hlmYNMwo40&t = 270s, last accessed on March 16, 2021.

6. The reportage can be viewed at https://www.youtube.com/watch?v = Zlyu7XYips4, last accessed on March 16, 2021.

7. Merleau-Ponty borrowed the notion of *Einfühlung* from Husserl and used it in its original German (see Moran 2015; Abram 2017).

8. Even though I adopt the analytical distinction between emotion and affect as it is defined by Brian Massumi (2002), I also take Sara Ahmed's warning seriously when she argues that a strict analytical distinction between affect and emotion "risks cutting emotions off from the lived experience of being and having a body" (Ahmed 2004, 39). While acknowledging the heuristic value of an analytical difference, I put the emphasis on the bodily nature of emotions and on the close affinity between senses, affects, and emotions.

9. The last sentence of Semra's statement was also cited in chapter 2, in a different context.

10. As González-Hidalgo and Zogrofos (2017) point out, emotions can also reproduce hegemonic relations of power and obstruct political action instead of enabling it.

CHAPTER 4

1. This chapter, like the previous one, draws mainly, but not exclusively, on the discourses and narratives of women. This does not mean, however, that men do not have place attachments and body memories that involve river waters, especially as children, and that they do not have strong place identities. They do. Women, however, tend to mention their body memories and personal histories and family heritage as tied to those memories more often while talking about their motivations to fight against hydropower projects. The reason why men do not talk about them as often as women do might be that they do not perceive such stories and narratives as "serious" (see chapter 2).

2. The ideal-typical nature of cosmopolitan subjects and thinned-out places aside, it might be that it is not the places that are thinning out, but our connections to those places, depending on the characteristics of our daily practices: "The thinning-out is primarily of the habitus linking places and selves—or more exactly, a replacement of one set of habitudes (more apt for lasting, lively engagements with robust places) with another habitudinal set (geared to leveled down places or 'sites')" (Casey 2001, 686; see also Strang 2005).

3. This is not to suggest that Turkey's rural landscapes are immune to spatial transformation. Still, the meanings we attach to spaces and places, the ways in which we experience and imagine them, shape our connections to them and motivate our political actions to defend them (Soja 1996; Dukpa et al. 2018).

4. As an initiative of the Turkish Association for Nature (*Doğa Derneği*) and Turkish Water Assembly, a meeting was organized at the European Parliament on November 20, 2010, at which representatives of many different local movements that fight against hydropower plants presented their cases and informed the EU representatives about the destructive effects of excessive hydropower development.

5. De la Cadena (2010, 354) cites Juan Oxa, whose identification with his home-place manifests the relationship between place attachment, place identity, and place-based struggles: "It is important to remember that this place is not where we are from; it is who we are. I am not from Huantura; I am Huantura."

6. The trend of "returning" is now statistically visible, as the Black Sea Region has begun to emerge as a migrant-receiving region, after being a migrant-sending region for decades. According to the data provided by the Turkish Statistical Institute, the Black Sea Region ranked high among receiving regions in 2018, with four major districts, Ordu, Rize, Giresun, and Trabzon, all among the top ten receiving

districts. See https://www.hurriyet.com.tr/galeri-iste-turkiyenin-en-cok-goc-alan-ve-goc-veren-illeri-ilk-siradaki-sehre-inanamayacaksiniz-41150661/3, last accessed on March 17, 2021.

7. And as long as rivers make places, they are crucial for our sense of place and place-bound identities, not only in Anatolia but also in many other parts of the world. Let us look, for example, at a verse that Murton (2012, 96) cites from the Hokianga area of northern New Zealand:

The mountain is Whakarongorua
The river is Utakuna
The sub-tribe is Te Honihoni
The people are Ngapuhi

In this verse, identity is defined in relation to the mountain and the river, which characterize the place.

8. Also referred to as "performative remembering" or "habitual body memory" (Casey 2000), this type of body memory is identified as "well-practiced patterns of movement and perception [that become] embodied as skills and capacities that we apply in our everyday lives as a matter of course" (Fuchs 2012, 10).

9. We are more concerned with the bodily memories of human beings here. But water bodies also keep the past in the present, and as we are all part of the planetary hydraulic cycle, the past goes through us constantly in the form of water. Astrida Neimanis (2017, 3) expresses this poetically: "Water extends embodiment in time—body, to body, to body . . . watery bodies are gestational milieus for another."

10. *Akıntıya Karşı* was made by a documentary collective (Umut Kocagöz, Özlem Işıl, Volkan Işıl, and Ezgi Akyol) in 2011 and is available with English subtitles at https://www.youtube.com/watch?v = kmj_JPuJxik, last accessed on March 17, 2021.

11. These stories and memories also underline the importance of rivers for providing a place for young people, men and women, to gather, socialize, and play.

12. Casey defines the "momentary" or "actual" body as "the lived body as it operates to meet the particular demands of a given moment" (Casey 2000, 335). The "customary" or "habitual" body, on the other hand, refers to our more general, pre-reflexive existence, which "serves to guarantee the actions of the momentary body" (Casey 2000).

13. Rivers can also remind people of much more painful hardships than those Cemal endured, such as the death of a loved one who drowned in the river. I have not come across such stories in my own research, probably because the streams of the villages I visited were not so large and deep, but I know that such stories exist, especially in Artvin, around the Çoruh River.

14. There are not many beaches on the Black Sea coast, and the inland region is very much disconnected from the sea due to the Black Sea Coastal Highway (D 010), which was opened in 2007. During the 1990s and early 2000s, a group of local activists and environmentalists organized a campaign to oppose the highway project, but they failed to prevent its construction along the coast. Cihan Eren, an environmentalist lawyer from Fındıklı whose name was associated with the antihighway campaign, was killed in 2005, just two days before the legal examination of the lawsuit regarding the project, of which he was the principal attorney.

CHAPTER 5

1. Both the distributive paradigm and the capabilities approach (e.g., Rawls 1971; Sen 2009; Nussbaum 2011) share this (moral) frame as reflected by concepts such as impartiality, fairness, duty, obligation, moral status, etc. Justice theorists such as Rainer Forst criticize this frame for being "pre-democratic," as "it accords priority to teleological values which are supposed to ground a just or good social order, where those who are subjected to this order do not feature in it as authors" (Forst 2014, 4).

2. Distributive justice, or social justice (the terms are often used interchangeably; see, e.g., Dobson 1998), is concerned with the distribution of goods, benefits, and burdens within society. John Rawls's seminal *A Theory of Justice* remains the main reference in the discussion on distributive justice since its publication in 1971. In it Rawls (1971, 8) defines social justice as "a standard whereby the distributive aspects of the basic structure of the society are to be assessed." His book is concerned with the formulation of generalizable, universal principles of organizing those distributive aspects, such as fairness and impartiality.

3. Nancy Fraser developed her tripartite theory over time, first combining the aspect of recognition with redistribution as issues of class and status hierarchy (Fraser 1997), and then integrating representation into her model (Fraser 2005). The formulation of recognition (see Young 1990; Taylor 1992; Honneth 1995) and representation as essential dimensions of justice has to do with the grassroots claims of justice in relation to identity, difference, and democracy, and the rise of social movements within which those claims are articulated (Yaka 2019b).

4. Recognition has been constructed as a main principle of justice in the writings of figures such as Iris Marion Young (1990) and Axel Honneth (1995). Honneth builds on Hegel's (2018 [1807]) and Mead's (1934) intersubjective conception of the self in conceptualizing recognition as the prerequisite of self-realization, self-respect, and self-esteem. Young, on the other hand, refers directly to politics of identity and difference in stressing the limits of the distributive paradigm, which, she argues, falls short of responding to injustice as

oppression. The critique of new social movements, directed at diverse forms of patriarchy, racism, colonialism, and heteronormativity, was certainly instrumental in the formulation of recognition as a main pillar of justice. As the issues of identity and difference come to the fore, misrecognition is identified as a form of oppression that inflicts as much harm on individuals and social groups as maldistribution does (Young 1990; Taylor 1992).

5. Ordinary political misrepresentation occurs when "political decision rules wrongly deny some of the included the chance to participate fully, as peers" (Fraser 2009, 18–19).

6. There are significant attempts to extend theories of recognition to include nature and the nonhuman world. David Schlosberg's discussion of an extended notion of recognition, which entails recognizing similarities between human and nonhuman worlds and the integrity of ecological systems, is especially worth mentioning here. However, the limits of "justice as recognition" become clear when Schlosberg turns to the established theories of recognition (which are either about individual dignity and status hierarchy or identity and difference; see Young 1990; Taylor 1992; Honneth 1995, 2004; Fraser and Honneth 2003) and tries to apply them to the nonhuman world. His application, based on the injurious misrecognition of nature, might facilitate a moral discussion on the idea of ecological justice, but it falls short of accounting for the relational nature of human and nonhuman life on Earth, for the very centrality of the nonhuman forms, ecologies, and natures for the socio-ecological existence of humans.

7. Subject pronouns are not gendered in Turkish. The single pronoun *o* is used where English would use *he*, *she*, and *it*.

8. See https://www.youtube.com/watch?v = Qr79ZtFQLYk, last accessed on June 6, 2022.

9. Earth practices are relations that involve Earth beings (other-than-human) that "enact respect and affect necessary to maintain the relational condition between humans and other-than-human beings" and "for which the dominant ontological distinction between humans and nature does not work" (de la Cadena 2010, 341).

10. Here it is also relevant to remember Deleuze, who claims that emotions and affects stand at the core of spiritual entities and abstract entities, as Stephanie Clare (2019, 24) recently reminded us.

11. This material and corporeal context in which cultural meanings, beliefs, and values are produced might be the very basis of "recurrent themes of meaning in relation to water" (Strang 2005, 93) that cut across the most diverse cultures and geographies.

12. The Achuar people are an Indigenous Amazonian community who live in the borderlands between Ecuador and Peru.

13. Many of Turkey's rural communities could be seen as indigenous, not in terms of being subjected to settler-state colonialism and violence (although we

can speak of settler colonialism in the context of Kurdish people and the Greek, Laz, and Hemshin peoples of the East Black Sea Region to some extent), but in the sense of "to be of a place," of being rooted in place for generations.

14. Even though certain environmental justice scholars such as Stacy Alaimo (2010, 2) demonstrate "the material interconnections between the human and more-than-human world," they have not discussed the conceptual implications of their significant work in formulating the "justice" part of environmental justice. The environmental justice literature has become more involved with Indigenous relational ontologies lately, but there is a tendency to reduce them to culture and cosmology, and thus to categorize the justice claims emerging from those relational ontologies and ethics under the banner of recognition.

CHAPTER 6

1. Despite its success in attracting public visibility and support, the struggle could not prevent the mining operations of Eurogold in Bergama. The movement managed to force the company to make substantial changes to the project, though, to limit the risk posed by cyanide leaching. See the Environmental Justice Atlas page on the Bergama Gold Mine at https://ejatlas.org/conflict/bergama-gold-mine-turkey, last accessed on March 18, 2021.

2. In the grip of repressive measures by the increasingly authoritarian Justice and Development Party government, most social movements and oppositional voices are struggling to survive. Grassroots environmentalisms, however, along with the women's movement, are an exception; they have managed to maintain their public visibility and efficacy under this authoritarian rule, as is manifested in the effective resistance against mining projects in Artvin Cerrattepe (EBR) and on Ida Mountain, located on the border of Balıkesir and Çanakkale provinces (Aegean Region) (see Gonenc 2016; MacDonald 2020).

3. "Thus there is an indivision of my body, of my body and the world, of my body and other bodies, and of other bodies between them" (Merleau-Ponty 2003, 279). Expressions of an "indivision" are not unique to the people of the East Black Sea Region, but shared by many rural and Indigenous communities around the world. In this sense, the case of the East Black Sea Region shows that the relational understandings of human and nonhuman existence that involve both ontological and epistemological aspects, and which are, more often than not, attributed to peculiar cultures and cosmologies, are "deeply embedded in the experience of everyday life" (Ingold 1999, 81).

4. This does not mean, of course, that abstract ideas and interests do not play a role in the realm of politics; it means, rather, that those ideas and interests are constituted within a world of bodily encounters. Such a conception of politics is also important for understanding contemporary social-environmental move-

ments and struggles, as they are not always motivated by instrumental goals or narrowly defined interests. If the wording of interests is essential to political language, then *interests* should be defined broadly, encompassing a wider world of experiences, encounters, connections, memories, ways of life, identities, etc. (see, e.g., Espeland 1998).

5. This is why claiming rights over natural entities and environments without owning them captures the core of contemporary struggles for and around the environmental commons.

References

Abram, David. 2017. *The Spell of the Sensuous: Perception and Language in a More-Than-Human World*. Twentieth Anniversary Edition. New York: Vintage.

Adaman, Fikret, Murat Arsel, and Bengi Akbulut. 2019. "Neoliberal Developmentalism, Authoritarian Populism, and Extractivism in the Countryside: The Soma Mining Disaster in Turkey." *Journal of Peasant Studies* 46 (3): 514–36.

Agarwal, Bina. 1992. "The Gender and Environment Debate: Lessons from India." *Feminist Studies* 18 (1): 119–58.

Agyeman, Julian, David Schlosberg, Luke Craven, and Caitlin Matthews. 2016. "Trends and Directions in Environmental Justice: From Inequity to Everyday Life, Community, and Just Sustainabilities." *Annual Review of Environment and Resources* 41: 321–40.

Åhäll, Linda. 2018. "Affect as Methodology: Feminism and the Politics of Emotion." *International Political Sociology* 12 (1): 36–52.

Ahlers, Rhodante. 2010. "Fixing and Nixing: The Politics of Water Privatization." *Review of Radical Political Economics* 42 (2): 213–30.

Ahlers, Rhodante, and Margreet Zwarteveen. 2009. "The Water Question in Feminism: Water Control and Gender Inequities in a Neo-Liberal Era." *Gender, Place and Culture* 16 (4): 409–26.

Ahmed, Sara. 2004. "Collective Feelings: Or, the Impressions Left by Others." *Theory, Culture & Society* 21 (2): 25–42.

Ahmed, Sara, and Jackie Stacey. 2001. *Thinking through the Skin*. London: Routledge.

Akbulut, Bengi, Murat Arsel, and Fikret Adaman. 2016. "Türkiye'de Kalkınmacılığı Yeniden Okumak: HES'ler ve Dönüşen Devlet-Toplum-Doğa İlişkileri [Re-Reading Developmentalism in Turkey: HPPs and Transforming State-Society-Nature Relations]." In *Sudan Sebepler: Türkiye'de Neoliberal Su-Enerji Politikaları ve Direnişler [Neoliberal Politics of Water-Energy in Turkey and Resistance]*, edited by Cemil Aksu, Sinan Erensü, and Erdem Evren, 291–312. İstanbul: İletişim Yayınları.

Akduman, Ismail. 2020. "TMMOB Giresun'daki Felaketin Nedenini Açıkladı: HES, Kaçak Yapılaşma, Yerel Yönetim Politikaları . . . [TMMOB Explains the Reasons of the Disaster: HEPP, Shanty Settlements, Policies of the Local Government . . .]." *Sözcü*, 23 August 2020. https://www.sozcu.com.tr/2020/gundem/tmmob-giresundaki-felaketin-nedenini-acikladi-hes-kacak-yapilasma-yerel-yonetim-politikalari-6002811/.

Aksu, Cemil, Sinan Erensü, and Erdem Evren. 2016. *Sudan Sebepler: Türkiye'de Neoliberal Su-Enerji Politikaları ve Direnişler [Neoliberal Politics of Water-Energy in Turkey and Resistance]*. İstanbul: İletişim Yayınları.

Aksu, Cemil, and Ramazan Korkut. 2017. *Ekoloji Almanağı: 2005-2016 [The Ecology Almanac: 2005-2016]*. İstanbul: Yeni İnsan Yayınevi.

Al-Saji, Alia. 2001. "Merleau-Ponty and Bergson: Bodies of Expression and Temporalities in the Flesh." *Philosophy Today* 45 (Supplement): 110–23.

Alaimo, Stacy. 2000. *Undomesticated Ground: Recasting Nature as Feminist Space*. Ithaca, NY: Cornell University Press.

———. 2010. *Bodily Natures Science, Environment, and the Material Self*. Bloomington: Indiana University Press.

Alaimo, Stacy, and Susan J. Hekman. 2008. *Material Feminisms*. Bloomington: Indiana University Press.

Alcoff, Linda, and Elizabeth Potter. 1993. *Feminist Epistemologies*. London: Routledge.

Altınay, Ayşe Gül. 2004. *The Myth of the Military Nation: Militarism, Gender, and Education in Turkey*. New York: Palgrave Macmillan.

Altman, Irwin, and Setha M. Low. 1992. *Place Attachment*. New York: Plenum Press.

Anderson, Ben, and Paul Harrison. 2006. "Questioning Affect and Emotion." *Area* 38 (3): 333–5.

———. 2010. *Taking-Place: Non-Representational Theories and Geography*. Burlington, VT: Ashgate.

Anderson, Elizabeth. 2000. "Feminist Epistemology and Philosophy of Science." In *The Stanford Encyclopedia of Philosophy*, edited by Edward N. Zalta. https://stanford.library.sydney.edu.au/archives/spr2014/entries/feminism-epistemology/.

Ash, James. 2017. "Visceral Methodologies, Bodily Style and the Non-Human." *Geoforum* 82: 206–7.

Ash, James, and Paul Simpson. 2016. "Geography and Post-Phenomenology." *Progress in Human Geography* 40 (1): 48–66.

Atiyas, Izak, Tamer Çetin, and Gürcan Gülen. 2012. *Reforming Turkish Energy Markets: Political Economy, Regulation and Competition in the Search for Energy Policy.* New York: Springer.

Avcı, Ömür, and Muhammet Kaçar. 2010. "Başbakan Erdoğan, İkizdere'de HES'i Hizmete Açarken Çevrecilere Çattı [Prime Minister Erdoğan Tilted at the Environmentalists While Opening the HEPP in İkizdere]." *Milliyet*, 8 November 2010. https://www.milliyet.com.tr/siyaset/basbakan-erdogan-ikizderede-hes-i-hizmete-acarken-cevrecilere-catti-1275291.

Babacan, Errol, Melehat Kutun, Ezgi Pınar, and Zafer Yılmaz, eds. 2021. *Regime Change in Turkey: Neoliberal Authoritarianism, Islamism and Hegemony.* London: Routledge.

Bachelard, Gaston. 1983. *Water and Dreams: An Essay on the Imagination of Matter.* Dallas: Pegasus Foundation.

Baird, Ian G., Bruce Shoemaker, and Kanokwan Manorom. 2015. "The People and Their River, the World Bank and Its Dam: Revisiting the Xe Bang Fai River in Laos." *Development and Change* 46 (5): 1080–1105.

Bakker, Karen. 2003. *An Uncooperative Commodity: Privatizing Water in England and Wales.* Oxford: Oxford University Press.

———. 2007. "The 'Commons' Versus the 'Commodity': Alter-Globalization, Anti-Privatization and the Human Right to Water in the Global South." *Antipode* 39 (3): 430–55.

Balibar, Etienne. 2012. "Justice and Equality: A Political Dilemma?" In *The Borders of Justice,* edited by Étienne Balibar, Sandro Mezzadra, and Ranabir Samaddar, 9–30. Philadelphia: Temple University Press.

Balibar, Etienne, Sandro Mezzadra, and Ranabir Samaddar, eds. 2012. *The Borders of Justice.* Philadelphia: Temple University Press.

Bannon, Bryan E. 2020. "The Bonding Properties of Water: Community, Urban River Restoration, and Non-Human Agency." In *The Wonder of Water: Lived Experience, Policy and Practice,* edited by I. L. Stefanovic, 151–70. Toronto: Toronto University Press.

Barad, Karen. 2007. *Meeting the Universe Halfway: Quantum Physics and the Entanglement of Matter and Meaning.* Durham, NC: Duke University Press.

Barbalet, Jack. 2002. "Introduction: Why Emotions Are Crucial." *The Sociological Review* 50 (2): 1–9.

Barbaras, Renaud. 2001. "Merleau-Ponty and Nature." *Research in Phenomenology* 31 (1): 22–38.

Başkaya, Şağdan, E. Başkaya, and Ayşegül Sarı. 2011. "The Principal Negative Environmental Impacts of Small Hydropower Plants in Turkey." *African Journal of Agricultural Research* 6 (14): 3284–90.

Baxter, Brain. 2005. *A Theory of Ecological Justice.* London: Routledge.

Baykan, Toygar Sinan. 2020. *The Justice and Development Party in Turkey: Populism, Personalism, Organization.* Cambridge: Cambridge University Press.

Bell, Shannon Elizabeth. 2013. *Our Roots Run Deep as Ironweed: Appalachian Women and the Fight for Environmental Justice.* Champaign: University of Illinois Press.

Bell, Shannon Elizabeth, and Yvonne A. Braun. 2010. "Coal, Identity, and the Gendering of Environmental Justice Activism in Central Appalachia." *Gender & Society* 24 (6): 794–813.

Bellér-Hann, Ildikó, and Chris Hann. 2000. *Turkish Region: State, Market & Social Identities on the East Black Sea Coast.* Oxford: School of American Research Press.

Benford, Robert D. 1997. "An Insider's Critique of the Social Movement Framing Perspective." *Sociological Inquiry* 67 (4): 409–30.

Benjamin, Andrew. 2015. *Towards a Relational Ontology: Philosophy's Other Possibility.* Albany, NY: SUNY Press.

Bennett, Jane. 2010. *Vibrant Matter: A Political Ecology of Things.* Durham, NC: Duke University Press.

Bennett, Vivienne. 1995. "Gender, Class, and Water: Women and the Politics of Water Service in Monterrey, Mexico." *Latin American Perspectives* 22 (2): 76–99.

Bennett, Vivienne, Sonia Dávila-Poblete, and Nieves Rico. 2005. *Opposing Currents: The Politics of Water and Gender in Latin America.* Pittsburgh, PA: University of Pittsburgh Press.

Bergson, Henri. 1988. *Matter and Memory.* Translated by Nancy Margaret Paul and W. Scott Palmer. New York: Urzone.

Berkes, Niyazi. 1964. *The Development of Secularism in Turkey.* Montreal: McGill University Press.

Bilgen, Arda. 2018. "A Project of Destruction, Peace, or Techno-Science? Untangling the Relationship between the Southeastern Anatolia Project (GAP) and the Kurdish Question in Turkey." *Middle Eastern Studies* 54 (1): 94–113.

Billington, David P., and Jackson Donald Conrad. 2006. *Big Dams of the New Deal Era: A Confluence of Engineering and Politics.* Norman: University of Oklahoma Press.

Bird-David, Nurit. 1999. "'Animism' Revisited: Personhood, Environment and Relational Epistemology." *Current Anthropology* 40 (1): 67–79.

Birgün. 2014. "Orman Bakanı Hala Anlamadı: Neden HESlere Karşısınız? [The Ministry of Forestry Still Does Not Get It: Why Are You Opposing the

HPPs?]," 10 December 2014. http://www.birgun.net/haber-detay/orman-bakani-hala-anlamadi-neden-hes-lere-karsisiniz-69902.html.

———. 2016. "Cumhurbaşkanı Erdoğan: 'Greenpeace'ci bizim Karadeniz'e hep bela oldu' [President Erdoğan: 'Greenpeace activists have always been trouble to our Black Sea']." birgun.net, 7 November 2016. https://www.birgun.net/haber/cumhurbaskani-erdogan-greenpeace-ci-bizim-karadeniz-e-hep-bela-oldu-134629.

Biryol, Uğur, ed. 2012. *Karardı Karadeniz [The Black Sea Has Darkened].* İstanbul: İletişim Yayınları.

Biryol, Ugur, ed. 2014. *Karadeniz'in kaybolan kimliğI [Disappearing Identity of the Black Sea].* Istanbul: Iletisim.

Blee, Kathleen M., and Verta Taylor. 2002. "Semi-Structured Interviewing in Social Movement Research." In *Methods of Social Movement Research,* edited by Bert Klandermans and Suzanne Staggenborg, 92–117. Social Movements, Protest and Contention Series, Volume 16. Minneapolis, London: University of Minnesota Press.

Boelens, Rutgerd. 2014. "Cultural Politics and the Hydrosocial Cycle: Water, Power and Identity in the Andean Highlands." *Geoforum* 57: 234–47.

———. 2015. *Water, Power and Identity: The Cultural Politics of Water in the Andes.* London: Routledge.

Boelens, Rutgerd, Jorge Armando Guevera Gil, and David Getches. 2010. *Out of the Mainstream: Water Rights, Politics and Identity.* London: Earthscan.

Bora, Tanıl. 2003. "Nationalist Discourses in Turkey." *The South Atlantic Quarterly* 102 (2): 433–51.

Bozdoğan, Sibel, and Resat Kasaba. 1997. *Rethinking Modernity and National Identity in Turkey.* Seattle: University of Washington Press.

Bozok, Mehmet. 2013. "Constructing Local Masculinities: A Case Study from Trabzon, Turkey." Unpublished PhD thesis, Middle East Technical University, Institute of Social Sciences. http://etd.lib.metu.edu.tr/upload/12615415/index.pdf.

Bozok, Mehmet, Nihan Bozok, and Meral Akbaş. 2016. "Bizim Dereyi Kim Çaldı?: Doğu Karadeniz'de Yaşlı Kadınlar ve Yaşlı Erkeklerin Doğa Anlatıları [Who Stole Our River? Old Women's and Old Men's Narratives of Nature in the East Black Sea]." In *Sudan Sebepler: Türkiye'de Neoliberal Su-Enerji Politikaları ve Direnişler [Neoliberal Politics of Water-Energy in Turkey and Resistance],* edited by Cemil Aksu, Sinan Erensü, and Erdem Evren. İstanbul: İletişim Yayınları.

Braidotti, Rosi. 1991. *Patterns of Dissonance: A Study of Women in Contemporary Philosophy.* London: Routledge.

———. 1994. *Nomadic Subjects: Embodiment and Sexual Difference in Contemporary Feminist Theory.* New York: Colombia University Press.

———. 2013. *The Posthuman.* Cambridge: Polity.

Braun, Bruce. 2004. "Querying Posthumanisms." *Geoforum* 35 (3): 269–73.

Braun, Yvonne A. 2008. "'How Can I Stay Silent?' One Woman's Struggles for Environmental Justice in Lesotho." *Journal of International Women's Studies* 10 (1): 5–20.

Bray, Abigail, and Claire Colebrook. 1998. "The Haunted Flesh: Corporeal Feminism and the Politics of (Dis) Embodiment." *Signs* 24 (1): 35–67.

Bromber, Katrin, Jeanne Féaux de la Croix, and Katharina Lange. 2014. "The Temporal Politics of Big Dams in Africa, the Middle East, and Asia: By Way of an Introduction." *Water History* 6 (4): 289–96.

Brown, Charles S., and Ted Toadvine. 2003. *Eco-Phenomenology: Back to the Earth Itself.* Albany: State University of New York Press.

Brown, Phil, and Faith I. T. Ferguson. 1995. "Making a Big Stink: Women's Work, Women's Relationships, and Toxic Waste Activism." *Gender & Society* 9 (2): 145–72.

Buckingham, Susan, and Rakibe Kulcur. 2009. "Gendered Geographies of Environmental Injustice." *Antipode* 41 (4): 659–83.

Buechler, Stephanie, and Anne-Marie S. Hanson. 2015. *A Political Ecology of Women, Water and Global Environmental Change.* London: Routledge.

Burawoy, Michael. 1998. "The Extended Case Method." *Sociological Theory* 16 (1): 4–33.

Butler, Judith. 1989. "Sexual Ideology and Phenomenological Description." In *The Thinking Muse: Feminism and Modern French Philosophy*, edited by Jeffner Allen and Iris Marion Young, 85–100. Bloomington: Indiana University Press.

———. 1990. *Gender Trouble: Feminism and the Subversion of Identity.* London: Routledge.

———. 2001. "Sexual Difference as a Question of Ethics." In *Bodies of Resistance: New Phenomenologies of Politics, Agency, and Culture*, edited by Laura Boyle, 59–77. Evanston, IL: Northwestern University Press.

———. 2005. "Merleau-Ponty and the Touch of Malebranche." In *The Cambridge Companion to Merleau-Pontyv*, edited by Taylor Non and Mark B. N. Hansen, 181–205. Cambridge: Cambridge University Press.

———. 2015. *Senses of the Subject.* New York: Fordham University Press.

Buttimer, Anne. 1976. "Grasping the Dynamism of Lifeworld." *Annals of the Association of American Geographers* 66 (2): 277–92.

———. 1980. "Home, Reach, and the Sense of Place". In *The Human Experience of Space and Place*, edited by Anne Buttimer and David Seamon. London: Routledge.

Caffentzis, George. 2010. "The Future of 'The Commons': Neoliberalism's 'Plan B' or the Original Disaccumulation of Capital?" *New Formations* 69: 23–41.

Çakıroğlu, Ali, and Mine Eder. 2001. "Domestic Concerns and the Water Conflict over the Euphrates-Tigris River Basin." *Middle Eastern Studies* 37 (1): 41–71.

Čapek, Stella M. 1993. "The 'Environmental Justice' Frame: A Conceptual Discussion and an Application." *Social Problems* 40 (1): 5–24.

Çapik, Mehmet, Ali Osman Yılmaz, and İbrahim Çavuşoğlu. 2012. "Present Situation and Potential Role of Renewable Energy in Turkey." *Renewable Energy* 46: 1–13.

Carman, Taylor. 2012. "Foreword." In *Phenomenology of Perception*, by Maurice Merleau-Ponty, translated by Donald A. Landes, vii–xvi. Oxon and New York: Routledge.

Casey, Edward S. 1993. *Getting Back into Place: Toward a Renewed Understanding of the Place-World*. Bloomington: Indiana University Press.

———. 2000. *Remembering: A Phenomenological Study*. Bloomington: Indiana University Press.

———. 2001. "Between Geography and Philosophy: What Does It Mean to Be in the Place-World?" *Annals of the Association of American Geographers* 91 (4): 683–93.

Castree, Noel. 2003. "Commodifying What Nature?" *Progress in Human Geography* 27 (3): 273–97.

———. 2004. "Differential Geographies: Place, Indigenous Rights and 'Local' Resources." *Political Geography* 23 (2): 133–67.

———. 2005. *Nature*. London: Routledge.

Chisholm, Dianne. 2008. "Climbing like a Girl: An Exemplary Adventure in Feminist Phenomenology." *Hypatia* 23 (1): 9–40.

Clare, Stephanie D. 2019. *Earthly Encounters. Sensation, Feminist Theory, and the Anthropocene*. Albany: State University of New York Press.

CNN Turk. 2010. "Başbakan Erdoğan HES'i Savundu Ama . . . [Prime Minister Erdoğan Advocated the HPPs But . . .]", 8 November 2010. http://www .cnnturk.com/2010/turkiye/08/11/basbakan.erdogan.hesi.savundu.ama /586384.0/index.html.

Çoban, Aykut. 2004. "Community-Based Ecological Resistance: The Bergama Movement in Turkey." *Environmental Politics* 13 (2): 438–60.

Code, Lorraine. 2006. *Ecological Thinking: The Politics of Epistemic Location*. Oxford: Oxford University Press.

Colebrook, Claire. 2017. "Materiality: Sex, Gender, and What Lies Beneath." In *The Routledge Companion to Feminist Philosophy*, edited by Alison Stone, Serene Khader, and Ann Garry, 194–206. London: Routledge.

Collard, Andrée, and Joyce Contrucci. 1989. *Rape of the Wild: Man's Violence against Animals and the Earth*. Bloomington: Indiana University Press.

Connolly, William E. 2017. *Facing the Planetary: Entangled Humanism and the Politics of Swarming*. Durham, NC: Duke University Press.

Coole, Diana. 2001. "Thinking Politically with Merleau-Ponty." *Radical Philosophy* 108, 17–28.

———. 2005. "Rethinking Agency: A Phenomenological Approach to Embodiment and Agentic Capacities." *Political Studies* 53 (1): 124–42.

———. 2007. *Merleau-Ponty and Modern Politics after Anti-Humanism.* Lanham, MD: Rowman & Littlefield.

Coole, Diana, and Samantha Frost. 2010. *New Materialisms: Ontology, Agency, and Politics.* Durham, NC: Duke University Press.

Coolsaet, Brendan. 2015. "Transformative Participation in Agrobiodiversity Governance: Making the Case for an Environmental Justice Approach." *Journal of Agricultural and Environmental Ethics* 28 (6): 1089–1104.

Coulthard, Glen. 2010. "Place against Empire: Understanding Indigenous Anti-Colonialism." *Affinities: A Journal of Radical Theory, Culture, and Action* 4 (2): 79–83.

Coulthard, Glen Sean. 2014. *Red Skin, White Masks: Rejecting the Colonial Politics of Recognition.* Minneapolis: University of Minnesota Press.

Coulthard, Glen, and Leanne Betasamosake Simpson. 2016. "Grounded Normativity/Place-Based Solidarity." *American Quarterly* 68 (2): 249–55.

Couto, Thiago B. A., and Julian D. Olden. 2018. "Global Proliferation of Small Hydropower Plants—Science and Policy." *Frontiers in Ecology and the Environment* 16 (2): 91–100.

Cruz-Torres, M. L., and Pamela McElwee. 2017. "Gender, Livelihoods and Sustainability: Anthropological Research." In *Routledge Handbook of Gender and Environment,* edited by Sherilyn MacGregor, 133–45. London: Routledge.

Culley, Marci R., and Holly L. Angelique. 2003. "Women's Gendered Experiences as Long-Term Three Mile Island Activists." *Gender & Society* 17 (3): 445–61.

Dahlberg, Karin, Nancy Drew, and Maria Nyström. 2001. *Reflective Lifeworld Research.* Lund, Sweden: Studentlitteratur.

Datta, Ranjan. 2015. "A Relational Theoretical Framework and Meanings of Land, Nature, and Sustainability for Research with Indigenous Communities." *Local Environment* 20 (1): 102–13.

De la Cadena, Marisol. 2010. "Indigenous Cosmopolitics in the Andes: Conceptual Reflections beyond 'Politics.'" *Cultural Anthropology* 25 (2): 334–70.

De Sousa, Ronald. 1990. *The Rationality of Emotion.* Cambridge, MA: MIT Press.

De Wolff, Kim, and Rina C. Faletti. 2022. "Introduction: Hydrohumanities." In *Hydrohumanities: Water Discourse and Environmental Futures,* edited by Kim De Wolff, Rina C. Faletti, and Ignacio López-Calvo, 1–15. Oakland: University of California Press.

Deane-Drummond, Celia. 2018. "Rivers at the End of the End of Nature: Ethical Trajectories of the Anthropocene Great Narrative." In *Rivers of the Anthropocene,* edited by J. M. Kelly, P. V. Scorpino, H. Berry, J. Syvitski, and M. Meybeck, 55–62. Oakland: University of California Press.

Della Porta, Donatella, ed. 2014. *Methodological Practices in Social Movement Research.* Oxford: Oxford University Press.

Della Porta, Donatella, and Mario Diani. 2006. *Social Movements: An Introduction.* Malden, MA: Blackwell.

Deniz, Dilşa. 2012. *Yol/Rê: Dersim İnanç Sembolizmi: Antropolojik Bir Yaklaşım [Yol/Rê: Dersim Belief Symbolism: An Anthropological Perspective].* İstanbul: İletişim.

———. 2016. "Dersim'de Su Kutsiyeti, Mizur/Munzur nehri İlişkisi, Anlamı ve Kapsamı ile Baraj/HES Projeleri [Sacred Water in Dersim, the Relationship of Mizur and Munzur River, Its Meaning and Scope and the Hydropower Projects]." In *Sudan Sebepler: Türkiye'de Neoliberal Su-Enerji Politikaları ve Direnişler [Neoliberal Politics of Water-Energy in Turkey and Resistance],* edited by Sinan Erensü, Cemil Aksu, and Erdem Evren, 177–97. İstanbul: İletişim.

Derrida, Jacques. 1992. "Force of Law: The 'Mystical Foundation of Authority.'" In *Deconstruction and the Possibility of Justice,* edited by D. Cornell, M. Rosenfeld, and D. G. Carlson, 3–67. London: Routledge.

Descola, Phillipe. 1996. *In the Society of Nature: A Native Ecology in Amazonia.* Cambridge: Cambridge University Press.

———. 2013. *Beyond Nature and Culture.* Chicago, IL: University of Chicago Press.

Di Chiro, Giovanna. 1992. "Defining Environmental Justice: Women's Voices and Grassroots Politics." *Socialist Review* 22 (4): 93–130.

Dobson, Andrew. 1998. *Justice and the Environment: Conceptions of Environmental Sustainability and Theories of Distributive Justice.* New York: Oxford University Press.

———. 2000. *Green Political Thought,* 3rd ed. London: Routledge.

Donohoe, Janet. 2014. *Remembering Places: A Phenomenological Study of the Relationship between Memory and Place.* Lanham, MD: Lexington Books.

———. 2020. "The Place of Water." In *The Wonder of Water: Lived Experience, Policy and Practice,* edited by Ingrid Leman Stefanovic, 79–90. Toronto: Toronto University Press.

Dörre, Klaus. 2012. "Landnahme, Das Wachstumsdilemma Und Die 'Achsen Der Ungleichheit.'" *Berliner Journal Für Soziologie* 22 (1): 101–28.

Drew, Georgina. 2017. *River Dialogues: Hindu Faith and the Political Ecology of Dams on the Sacred Ganga.* Tuscon, AZ: University of Arizona Press.

196   REFERENCES

Dukpa, Rinchu Doma, Deepa Joshi, and Rutgerd Boelens. 2018. "Hydropower Development and the Meaning of Place. Multi-Ethnic Hydropower Struggles in Sikkim, India." *Geoforum* 89: 60–72.
Eaton, Heather, and Lois Ann Lorentzen, eds. 2003. *Ecofeminism and Globalization: Exploring Culture, Context, and Religion*. Lanham, MD.: Rowman & Littlefield.
Eberlein, Christine, Heike Drillisch, Ercan Ayboğa, and Thomas Wenidoppler. 2010. "The Ilisu Dam in Turkey and the Role of the Export Credit Agencies and NGO Networks." *Water Alternatives* 3 (2): 291–312.
Edenhofer, Ottmar, Ramón Pichs Madruga, Youba Sokona, Programa de las Naciones Unidas para el Medio Ambiente, Organización Meteorológica Mundial, Intergovernmental Panel on Climate Change, and Working Group III. 2012. *Renewable Energy Sources and Climate Change Mitigation: Special Report of the Intergovernmental Panel on Climate Change*. New York: Cambridge University Press.
Edgeworth, Matt, and Jeffrey Benjamin. 2018. "What Is a River? The Chicago River as a Hyperobject." In *Rivers of the Anthropocene*, by J. M. Kelly, P. V. Scorpino, H. Berry, J. Syvitski, and M. Meybeck, 162–75. Oakland: University of California Press.
EMO (Chamber of Electrical Engineers). 2011. "Doğu Karadeniz Bölgesi HES Teknik Gezisi Raporu [East Black Sea Region HEPP Technical Visit Report]." Ankara: EMO—Chamber of Electrical Engineers. http://www .emo.org.tr/ekler/45a43a1706a8faf_ek.pdf?tipi=1&turu=X&sube=0.
Erensü, Sinan. 2016a. "Hareket, Devlet ve Sermaye Arasında Bir İmkan Olarak Hukuk: Fevzi Özlüer Ile Söyleşi [Law as a Possibility between the Movement, the State and the Capital: Interview with Fevzi Özlüer]." In *In Sudan Sebepler: Türkiye'de Neoliberal Su-Enerji Politikaları ve Direnişler [Neoliberal Politics of Water-Energy in Turkey and Resistance]*, edited by Cemil Aksu, Erdem Evren, and Sinan Erensü, 451–69. İstanbul: İletişim.
———. 2016b. "Ölçüyü Doğru Koymanın Hukuku: Yakup Okumuşoğlu Ile Söyleşi [The Law of Putting the Measure Right: Interview with Yakup Okumuşoğlu]." In *Sudan Sebepler: Türkiye'de Neoliberal Su-Enerji Politikaları ve Direnişler [Neoliberal Politics of Water-Energy in Turkey and Resistance]*, edited by Cemil Aksu, Erdem Evren, and Sinan Erensü, 433–50. İstanbul: İletişim.
———. 2018. "Powering Neoliberalization: Energy and Politics in the Making of a New Turkey." *Energy Research & Social Science* 41: 148–57.
Erickson, Edward J. 2021. *The Turkish War of Independence: A Military History, 1919–1923*. Santa Barbara, CA: Praeger.
Eryılmaz, Çağrı. 2018. "Türkiye'de Çevreci Örgütlerin Dönüşümü: Merkezi Profesyonel Lobici Örgütler ve Yerelde Gönüllü Protestocular." *Ankara Üniversitesi SBF Dergisi* 73 (1): 49–76.

Escobar, Arturo. 2001. "Culture Sits in Places: Reflections on Globalism and Subaltern Strategies of Localization." *Political Geography* 20 (2): 139–74.

———. 2011. "Sustainability: Design for the Pluriverse." *Development* 54 (2): 137–40.

Esen, Berk, and Sebnem Gumuscu. 2018. "Building a Competitive Authoritarian Regime: State–Business Relations in the AKP's Turkey." *Journal of Balkan and Near Eastern Studies* 20 (4): 349–72.

Espeland, Wendy Nelson. 1998. *The Struggle for Water: Politics, Rationality, and Identity in the American Southwest.* Chicago: University of Chicago Press.

Esposito, Roberto. 2015. *Persons and Things: From the Body's Point of View.* Cambridge: Polity Press.

Evans, Fred, and Leonard Lawlor. 2000. *Chiasms: Merleau-Pontys Notion of Flesh.* Albany: State University of New York Press.

Evren, Erdem. 2014. "The Rise and Decline of an Anti-Dam Campaign: Yusufeli Dam Project and the Temporal Politics of Development." *Water History* 6 (4): 405–19.

———. 2022. *Bulldozer Capitalism: Accumulation, Ruination, and Dispossession in Northeastern Turkey.* Oxford and New York: Berghahn Books.

Fahim, Hussein M. 1981. *Dams, People, and Development: The Aswan High Dam Case.* New York: Pergamon Press.

Federici, Silvia. 2012. *Revolution at Point Zero: Housework, Reproduction, and Feminist Struggle.* Oakland, CA: PM Press.

Fielding, Helen A. 2017. "A Feminist Phenomenology Manifesto." In *Feminist Phenomenology Futures,* edited by Helen A. Fielding and D. E. Olkowski. Bloomington: Indiana University Press.

Fielding, Helen, and Dorothea Olkowski, eds. 2017. *Feminist Phenomenology Futures.* Bloomington: Indiana University Press.

Forst, Rainer. 2014. *Justification and Critique: Towards a Critical Theory of Politics.* Translated by C. Cronin. Cambridge, UK; Malden, MA: Polity Press.

Foster, John B. 1999. "Marx's Theory of Metabolic Rift: Classical Foundations for Environmental Sociology." *American Journal of Sociology* 105 (2): 366–405.

Foster, John B., and Brett Clark. 2018. "The Expropriation of Nature." *Monthly Review* 69 (10): 1–27.

Foster, John B., Brett Clark, and Richard York. 2010. *The Ecological Rift: Capitalism's War on the Earth.* New York: Monthly Review Press.

Fox, Warwick. 1990. *Toward a Transpersonal Ecology: Developing New Foundations For Environmentalists.* Boston: Shambhala Press.

Fox, Warwick, University of Tasmania, Board of Environmental Studies, University of Tasmania, and Centre for Environmental Studies. 1986.

Approaching Deep Ecology: A Response to Richard Sylvan's Critique of Deep Ecology. Hobart: Board of Environmental Studies, University of Tasmania.

Franklin, Sarah, Celina Lury, and Jackie Stacey. 2000. *Global Nature, Global Culture.* London: SAGE.

Fraser, Nancy. 1997. *Justice Interruptus: Critical Reflections on the "Postsocialist" Condition.* London: Routledge.

———. 2005. "Reframing Justice in a Globalizing World." *New Left Review* 36: 69–88.

———. 2009. *Scales of Justice: Reimagining Political Space in a Globalizing World.* New York: Colombia University Press.

———. 2014. "Can Society Be Commodities All the Way Down? Post-Polanyian Reflections on Capitalist Crisis." *Economy and Society* 43 (4): 541–58.

———. 2016. "Expropriation and Exploitation in Racialized Capitalism: A Reply to Michael Dawson." *Critical Historical Studies* 3 (1): 163–78.

Fraser, Nancy, and Axel Honneth. 2003. *Redistribution or Recognition? A Political-Philosophical Exchange.* London: Verso.

Fuchs, Thomas. 2012. "The Phenomenology of Body Memory." In *Body Memory, Metaphor and Movement,* edited by Sabine C. Koch, Thomas Fuchs, and Michela Summa. Amsterdam: John Benjamins.

———. 2013. "The Phenomenology of Affectivity." In *The Oxford Handbook of Philosophy and Psychiatry,* edited by K. W. M. Fulford, Martin Davies, Richard Gipps, George Graham, John Z. Sadler, Giovanni Stanghellini, and Tim Thornton, 612–31. Oxford: Oxford University Press.

———. 2018. *Ecology of the Brain: The Phenomenology and Biology of the Embodied Mind.* Oxford: Oxford University Press.

Gaard, Greta, ed. 1993. *Ecofeminism: Women, Animals, Nature.* Philadelphia: Temple University Press.

———. 2017. "Posthumanism, Ecofeminism, and Inter-Species Relations." In *Routledge Handbook of Gender and Environment,* edited by Sherilyn MacGregor, 115–29. London: Routledge.

Gamson, William A. 1992. *Talking Politics.* New York: Cambridge University Press.

Garland, Anne W. 1988. *Women Activists: Challenging the Abuse of Power.* New York: Feminist Press at the City University of New York.

Gatens, Moira. 1988. "Toward a Feminist Philosophy of the Body." In *Crossing Boundaries: Feminisms and the Critique of Knowledges,* edited by E. Caine, E. Grosz, and M. De Lepervanche, 59–70. Sydney: Allen and Unwin.

———. 1996. *Imaginary Bodies: Ethics, Power and Corporeality.* London: Routledge.

Gazioğlu, Elif. 2014. "Doğu Karadeniz Bölgesinin Toplumsal Cinsiyet Rejimi ve Kadınların Toplumsal Konumları [Gender Regime in the East Black Sea

Region and the Social Status of Women]." *Karadeniz Araştırmaları* 40: 95–108.

Gelles, P. H. 2010. "Cultural Identity and Indigenous Water Rights in the Andean Highlands." In *Out of the Mainstream*, edited by R. Boelens, D. H. Getches, and J. A. Guevera Gil, 137–62. London: Earthscan.

Gergan, Mabel Denzin. 2015. "Animating the Sacred, Sentient and Spiritual in Post-Humanist and Material Geographies." *Geography Compass* 9 (5): 262–75.

Ghosh, Jayati. 2010. "The Unnatural Coupling: Food and Global Finance." *Journal of Agrarian Change* 10 (1): 72–86.

Gibbons, Fiachra, and Lucas Moore. 2011. "Turkey's Great Leap Forward Risks Cultural and Environmental Bankruptcy." Guardian, 29 April 2011. http:// www.theguardian.com/world/2011/may/29/turkey-nuclear-hydro-power-development.

Gleick, Peter H. 1998. "The Human Right to Water." *Water Policy* 1 (5): 487–503.

Goldsmith, Edward, and Nicholas Hildyard, eds. 1984. *The Social and Environmental Effects of Large Dams: Volume 1 Overview*. Camelford, Cornwall, UK: Wadebridge Ecological Centre.

———, eds. 1986. *The Social and Environmental Effects of Large Dams: Volume 2 Case Studies*. Camelford, Cornwall, UK: Wadebridge Ecological Centre.

Gonenc, Defne. 2016. "Turkey's Largest Environmental Legal Case." OpenDemocracy. 8 October 2016. https://www.opendemocracy.net/en/turkey-s-largest-environmental-legal-case/.

Göner, Özlem. 2017. *Turkish National Identity and Its Outsiders: Memories of State Violence in Dersim*. London: Routledge.

González-Hidalgo, Marien, and Christos Zografos. 2017. "How Sovereignty Claims and 'Negative' Emotions Influence the Process of Subject-Making: Evidence from a Case of Conflict over Tree Plantations from Southern Chile." *Geoforum* 78: 61–73.

Goodwin, Jeff, James M. Jasper, and Francesca Polletta, eds. 2009. *Passionate Politics: Emotions and Social Movements*. Chicago: University of Chicago Press.

Gould, Deborah Bejosa. 2009. *Moving Politics: Emotion and ACT UP's Fight against AIDS*. Chicago: University of Chicago Press.

Gould, Kenneth Alan, Allan Schnaiberg, and Adam S Weinberg. 1996. *Local Environmental Struggles: Citizen Activism in the Treadmill of Production*. Cambridge: Cambridge University Press.

Gregg, Melissa, and Gregory J. Seigworth, eds. 2010. *The Affect Theory Reader*. Durham, NC: Duke University Press.

Griffin, Susan. 1978. *Woman and Nature: The Roaring Inside Her*. New York: Harper & Row.

Grosz, Elizabeth. 1987. "Notes Towards a Corporeal Feminism." *Australian Feminist Studies* 2 (5): 1–16.

———. 1993. "Bodies and Knowledges: Feminism and the Crisis of Reason." In *Feminist Epistemologies*, edited by Linda Alcoff and Elizabeth Potter, 187–216. London: Routledge.

———. 1994. *Volatile Bodies: Toward a Corporeal Feminism.* Bloomington: Indiana University Press.

Guattari, Felix. 2000. *Three Ecologies.* London: Athlone Press.

Gudynas, Eduardo. 2011. "Buen Vivir: Today's Tomorrow." *Development* 54 (4): 441–47.

Guenther, Lisa. 2013. *Solitary Confinement: Social Death and Its Afterlives.* Minneapolis: University of Minnesota Press.

Gunderson, Ryan. 2017. "Commodification of Nature." In *International Encyclopedia of Geography: People, the Earth, Environment and Technology*, edited by D. Richardson, N. Castree, M. F. Goodchild, A. Kobayashi, R. A. Marston, and W. Liu. Hoboken, NJ: AAG and Wiley.

Güneş, Şule. 2020. "Environmental Impact Assessment in Turkey: A Principal Environmental Management Tool." In *Environmental Law and Policies in Turkey*, edited by Z. Savaşan and V Sümer, Vol. 1: 83–97. Cham: Springer.

Habermas, Jürgen. 1987. *Theory of Communicative Action, Volume Two: Lifeworld and System, A Critique of Functionalist Reason.* Translated by Thomas A. McCarthy. Boston: Beacon Press.

Hamsici, Mahmut. 2010. *Dereler ve Isyanlar.* Ankara: Nota Bene.

Haraway, Donna. 1988. "Situated Knowledges: The Science Question in Feminism and the Privilege of Partial Perspective." *Feminist Studies* 14 (3): 575–99.

———. 1991. *Simians, Cyborgs, and Women: The Reinvention of Nature.* London: Routledge.

———. 2008. *When Species Meet.* Minneapolis: University of Minnesota Press.

Harcourt, Wendy, and Arturo Escobar, eds. 2005. *Women and the Politics of Place.* Bloomfield, CT: Kumarian Press.

Harcourt, Wendy, and L. Nelson Ingrid. 2015. *Practising Feminist Political Ecologies: Moving Beyond the "Green Economy."* London: Zed Books.

Harding, Sandra G., ed. 1987. *Feminism and Methodology: Social Science Issues.* Bloomington: Indiana University Press.

———. 2004. *The Feminist Standpoint Theory Reader: Intellectual and Political Controversies.* London: Routledge.

Harding, Sandra G., and Kathryn Norberg. 2005. "New Feminist Approaches to Social Science Methodologies: An Introduction." *Signs* 30 (4): 2009–15.

Hardt, Michael, and Antonio Negri. 2009. *Commonwealth.* Cambridge, MA: Belknap Press of Harvard University Press.

———. 2017. *Assembly.* Oxford: Oxford University Press.

Harris, Leila M. 2002. "Water and Conflict Geographies of the Southeastern Anatolia Project." *Society & Natural Resources* 15 (8): 743–59.

———. 2009. "Gender and Emergent Water Governance: Comparative Overview of Neoliberalized Natures and Gender Dimensions of Privatization, Devolution and Marketization." *Gender, Place and Culture* 16 (4): 387–408.

Harrison, Paul. 2007. "The Space between Us: Opening Remarks on the Concept of Dwelling." *Environment and Planning D: Society and Space* 25 (4): 625–47.

Harvey, David. 2003. *The New Imperialism*. Oxford: Oxford University Press.

Hay, Robert. 1998. "Sense of Place in Developmental Context." *Journal of Agricultural and Environmental Psychology* 18 (1): 5–29.

Hegel, Georg Wilhelm Friedrich. 2018. *The Phenomenology of Spirit*. Translated by M. J. Inwood. Oxford: Oxford University Press.

Heidegger, Martin. 1971. "Building, Dwelling, Thinking." In *Poetry, Language, Thought,* translated by Albert Hofstadter. New York: Harper Colophon Books.

———. 1996. *Hölderlin's Hymn "The Ister."* Translated by William McNeill and Julia Davis. Bloomington: Indiana University Press.

Held, Virginia. 2005. *Ethics of Care: Personal, Political, and Global.* Oxford: Oxford University Press.

Henighan, Stephen, and Candace Johnson, eds. 2018. *Human and Environmental Justice in Guatemala.* Toronto: University of Toronto Press.

Heynen, Nik, and Paul Robbins. 2005. "The Neoliberalization of Nature: Governance, Privatization, Enclosure and Valuation." *Capitalism, Nature, Socialism* 16 (1): 5–8.

Hidalgo, M. Carmen, and Bernardo Hernández. 2001. "Place Attachment: Conceptual and Empirical Questions." *Journal of Agricultural and Environmental Psychology* 21 (3): 273–81.

Highmore, Ben. 2010. "Bitter After Taste: Affect, Food and Social Aesthetics." In *The Affect Theory Reader,* edited by Melissa Gregg and Gregory J. Seigworth, 118–37. Durham, NC: Duke University Press.

Hochschild, Arlie Russell. 1998. "The Sociology of Emotion as a Way of Seeing." In *Emotions in Social Life: Critical Themes and Contemporary Issues,* edited by Gillian Bendelow and Simon J. Williams, 3–15. London: Routledge.

Hommes, Lena, Rutgerd Boelens, and Harro Maat. 2016. "Contested Hydrosocial Territories and Disputed Water Governance: Struggles and Competing Claims over the Ilisu Dam Development in Southeastern Turkey." *Geoforum* 71: 9–20.

Honneth, Axel. 1995. *The Struggle for Recognition: The Moral Grammar of Social Conflicts.* London: Polity Press.

———. 2004. "Recognition and Justice Outline of a Plural Theory of Justice." *Acta Sociologica* 47 (4): 351–64.

Howes, David. 2005. "Introduction: Empires of the Senses." In *Empire of Senses: The Sensual Culture Reader*, edited by David Howes. London: Bloomsbury.

Ihde, Don. 1990. *Technology and the Lifeworld: From Garden to Earth*. Bloomington: Indiana University Press.

———. 1993. *Postphenomenology: Essays in the Postmodern Context*. Evanston, IL: Northwestern University Press.

İnal, Onur, and Ethemcan Turhan, eds. 2020. *Transforming Socio-Natures in Turkey: Landscapes, State and Environmental Movements*. London: Routledge.

Inglehart, Ronald. 1990. *Culture Shift in Advanced Industrial Society*. Princeton, NJ: Princeton University Press.

Ingold, Tim. 1999. "Comment on '"Animism" Revisited: Personhood, Environment, and Relational Epistemology' by N Bird-David." *Current Anthropology* 40 (1): 81–82.

———. 2000. *The Perception of the Environment: Essays on Livelihood, Dwelling, and Skill*. London: Routledge.

———. 2012. "Towards an Ecology of Materials." *Annual Review of Anthropology* 41: 427–42.

IPCC (Intergovernmental Panel on Climate Change). 2011. "Hydropower." In *Special Report on Renewable Energy Sources and Climate Change Mitigation*, edited by Ottmar Edenhofer. Cambridge: Cambridge University Press.

Irigaray, Luce. 1993. *An Ethics of Sexual Difference*. Ithaca, NY: Cornell University Press.

Işıl, Özlem, and Özlem Arslan. 2014. "Kadınlığın Su Hali: Hes Direnişindeki Karadenizli Kadın Temsilleri Üzerine Bir Deneme [Liquid Form of Womenhood: An Essay on the Representations of Black Sea Women in the Context of the HPP Resistance]." *Kültür ve Siyasette Feminist Yaklaşımlar [Feminist Approaches in Culture and Politics]* 22: 40–48.

Işlar, Mine. 2012. "Privatised Hydropower Development in Turkey: A Case of Water Grabbing." *Water Alternatives* 5 (2): 376–91.

Jasper, James M. 2011. "Emotions and Social Movements: Twenty Years of Theory and Research." *Annual Review of Sociology* 37: 285–303.

Jenkins, Katy. 2015. "Unearthing Women's Anti-Mining Activism in the Andes: Pachamama and the 'Mad Old Women.'" *Antipode* 47 (2): 442–60.

Jessop, Bob. 2012. "Economic and Ecological Crises: Green New Deals and No-Growth Economies." *Development* 55: 17–24.

Jewett, Chas, and Mark Garavan. 2019. "Water Is Lif: An Indigenous Perspective from a Standing Rock Water Protector." *Community Development Journal* 54 (1): 42–58.

Jones, Owain, and Joanne Garde-Hansen. 2012a. "Identity." In *Geography and Memory: Explorations in Identity, Place and Becoming*, edited by Owain Jones and Joanne Garde-Hansen, 19–23. New York: Palgrave Macmillan.

———. 2012b. "Introduction." In *Geography and Memory: Explorations in Identity, Place and Becoming*, edited by Owain Jones and Joanne Garde-Hansen, 1–18. New York: Palgrave Macmillan.

———. 2012c. "Place." In *Geography and Memory: Explorations in Identity, Place and Becoming*, edited by Owain Jones and Joanne Garde-Hansen, 85–90. New York: Palgrave Macmillan.

Jongerden, Joost, ed. 2022. *The Routledge Handbook of Contemporary Turkey*. London: Routledge.

Kaçar, Muhammet. 2017. "Arılı Vadisinde Yola Oturan Kadınlar Bilirkişi Heyetinin Yolunu Kesti [The Women Block the Road in the Arılı Valley]." *Hürriyet*, 18 July 2017. https://www.hurriyet.com.tr/arili-vadisinde-yola-oturan-kadinlar-bilirkisi-40523696.

Kaika, Maria. 2006. "Dams as Symbols of Modernization: The Urbanization of Nature between Geographical Imagination and Materiality." *Annals of the Association of American Geographers* 92 (2): 276–301.

Kalender, Aytekin, and Selçuk Başar. 2018. "'Derelerin Anası' Arılı Vadisinde Nöbet Tutuyor [The Mother of the Rivers/Streams Keeps Guard at the Arılı Valley]." *T24 News*, 23 January 2018. https://t24.com.tr/haber/derelerin-anasi-arili-vadisinde-nobet-tutuyor,542478.

Kandiyoti, Deniz. 1988. "Bargaining with Patriarchy." *Gender & Society* 2 (3): 274–90.

———. 1991. "Islam and Patriarchy: A Comparative Perspective." In *Women in Middle Eastern History*, edited by N. R. Keddie and B. Baron. New Haven, CT: Yale University Press.

Karaçimen, Elif, and Ekin Değirmenci. 2019. "Kuşaktan Kuşağa Çay Tarımında Kadın Emeği [Women's Labor in the Tea Agriculture from One Generation to the Other]." In *Aramızda Kalmasın: Kır, Kent ve Ötesinde Toplumsal Cinsiyet [Not to Keep Between Us: Gender in and Beyond the Country and the City]*, edited by Özlem Şendeniz, 69–80. Fındıklı: Aramızda Toplumsal Cinsiyet Derneği Yayınları. https://aramizda.org.tr/wp-content/uploads/2020/02/Aramızda_Kalmasın.pdf.

Kasapoğlu, Çağıl. 2013. "100 Kadın: Karadeniz'in Dereleri İçin Direnen Kadınları [100 Women: The Black Sea Women Resisting for Their Rivers]." BBC Turkish, 25 October 2013. http://www.bbc.com/turkce/haberler/2013/10/131025_findikli_kadinlar_kasapoglu.

Kaygusuz, Özlem. 2018. "Authoritarian Neoliberalism and Regime Security in Turkey: Moving to an 'Exceptional State' under AKP." *South European Society and Politics* 23 (2): 281–302.

Kelly, Jason M., Philip Scarpino, Helen Berry, James Syvitski, and Michel
    Meybeck, eds. 2017. *Rivers of the Anthropocene.* Oakland: University of
    California Press.
Kepenek, Evrim. 2014. "Karadeniz'de Adım Adım Ekolojik Isyan [Step by Step
    Ecological Rebellion in the Black Sea]." *Özgür Gündem,* 24 June 2014.
Kerslake, Celia, Kerem Öktem, and Philip Robins, eds. 2010. *Turkey's Engage-
    ment with Modernity: Conflict and Change in the Twentieth Century.*
    London: Palgrave Macmillan.
Khagram, Sanjeev. 2004. *Dams and Development: Transnational Struggles for
    Water and Power.* Ithaca, NY: Cornell University Press.
King, Marcus Dubois, ed. 2000. *Water and Conflict in the Middle East.* Oxford:
    Oxford University Press.
Kinkaid, Eden. 2021. "Is Post-Phenomenology a Critical Geography? Subjectiv-
    ity and Difference in Post-Phenomenological Geographies." *Progress in
    Human Geography* 45 (2): 298–316.
Kirby, Vicki. 1997. *Telling Flesh: The Substance of the Corporeal.* London:
    Routledge.
Klandermans, Bert, and Suzanne Staggenborg. 2002. *Methods of Social
    Movement Research.* Minneapolis: University of Minnesota Press.
Klein, Naomi. 2014. *This Changes Everything: Capitalism vs. the Climate.* New
    York: Simon & Schuster.
Klingensmith, D. 2007. *"One Valley and a Thousand": Dams, Nationalism, and
    Development.* New Delhi: Oxford University Press.
Kömürcü, Murat İ., and Adem Akpınar. 2010. "Hydropower Energy versus
    Other Energy Sources in Turkey." *Energy Sources: Part B: Economics,
    Planning, and Policy* 5 (2): 185–98.
Köse, Hilal. 2016. "Sarı Yazmalı Halime Çakmak: Hava, Su, Toprak . . . Ötesi
    Var Mı?' [Halime Çakmak with the Yellow Scarf: Air, Water, Soil . . . Is
    There Anything Beyond That?]". Cumhuriyet, 3 June 2016. https://www
    .cumhuriyet.com.tr/haber/sari-yazmali-halime-cakmak-hava-su-toprak-
    otesi-var-mi-493114.
Krause, Franz, and Veronica Strang. 2016. "Thinking Relationships Through
    Water." *Society & Natural Resources* 29 (6): 633–38.
Krauss, Celene. 1993. "Women and Toxic Waste Protests: Race, Class and
    Gender as Resources of Resistance." *Qualitative Sociology* 16 (3): 247–62.
———. 1998. "Challenging Power: Toxic Waste Protests and the Politicization of
    White, Working-Class Women." In *Community Activism and Feminist
    Politics: Organizing across Race, Class and Gender,* edited by Nancy A.
    Naples, 129–50. London: Routledge.
Kruks, Sonia. 2001. *Retrieving Experience: Subjectivity and Recognition in
    Feminist Politics.* Ithaca, NY: Cornell University Press.

Kurasawa, Fuyuki. 2007. *The Work of Global Justice: Human Rights as Practices*. Cambridge: Cambridge University Press.

Kurdoğlu, Oğuz. 2016. "Expert-Based Evaluation of the Impacts of Hydropower Plant Construction on Natural Systems in Turkey." *Energy & Environment* 27 (6–7): 690–703.

Kurtz, H. E. 2007. "Gender and Environmental Justice in Louisiana: Blurring the Boundaries of Public and Private Spheres." *Gender, Place and Culture* 14 (4): 409–26.

LaDuke, Winona. 1999. *All Our Relations: Native Struggles for Land and Life*. Cambridge, MA: South End Press.

Lahiri-Dutt, Kuntala. 2015. "The Silent (and Gendered) Violence: Understanding Water Access in Mining Areas." In *A Political Economy of Women, Water and Global Environmental Change*, edited by Stephanie Buechler and Anne-Marie Hanson, 38–57. London: Routledge.

Landes, Donald A. 2012. "Translator's Introduction." In *Phenomenology of Perception*, edited by M. Merleau-Ponty, 30–51. London: Routledge.

Larsen, Soren C., and Jay T. Johnson. 2012a. "In between Worlds: Place, Experience, and Research in Indigenous Geography." *Journal of Cultural Geography* 29 (1): 1–13.

———. 2012b. "Toward an Open Sense of Place: Phenomenology, Affinity, and the Question of Being." *Annals of the Association of American Geographers* 102 (3): 632–46.

———. 2017. *Being Together in Place: Indigenous Coexistence in a More-than-Human World*. Minneapolis: University of Minnesota Press.

Latour, Bruno. 1993. *We Have Never Been Modern*. Cambridge, MA: Harvard University Press.

———. 2004. *Politics of Nature: How to Bring the Sciences into Democracy*. Cambridge, MA: Harvard University Press.

———. 2007. *Reassembling the Social: An Introduction to Actor-Network-Theory*. Oxford: Oxford University Press.

Law, John, and Annemarie Mol. 1995. "Notes on Materiality and Sociality." *Sociological Review* 43 (2): 274–94.

Le Breton, David. 2006. *Sensing the World: An Anthropology of the Senses*. Translated by Carmen Ruschiensky. London: Routledge.

Lee, Yuen-ching. 2014. "Water Power: The 'Hydropower Discourse' of China in an Age of Environmental Sustainability." *ASIANetwork Exchange: A Journal for Asian Studies in the Liberal Arts* 21 (1): 1–10.

Lefort, Claude. 2012. "Maurice Merleau-Ponty." In *Phenomenology of Perception*, by M. Merleau-Ponty, 17–29. London: Routledge.

Lennon, Kathleen, and Margaret Whitford. 1994. *Knowing the Difference: Feminist Perspectives in Epistemology*. London: Routledge.

Leopold, Aldo. 1949. *A Sand County Almanac*. Oxford: Oxford University Press.

Levin, Sam. 2016. "At Standing Rock, Women Lead Fight in Face of Mace, Arrests and Strip Searches." Guardian, 11 April 2016. https://www .theguardian.com/us-news/2016/nov/04/dakota-access-pipeline-protest-standing-rock-women-police-abuse.

Lewicka, Maria. 2008. "Place Attachment, Place Identity, and Place Memory: Restoring the Forgotten City Past." *Journal of Environmental Psychology* 28 (3): 209–31.

Lingis, Alphonso. 1968. "Translator's Preface." In *The Visible and the Invisible*, by M. Merleau-Ponty. Evanston, IL: Northwestern University Press.

Lloyd, Kate, Sarah Wright, Sandie Suchet-Pearson, Laklak Burarrwanga, and Bawaka Country. 2012. "Reframing Development through Collaboration: Towards a Relational Ontology of Connection in Bawaka, North East Arnhem Land." *Third World Quarterly* 33 (6): 1075-94.

Longhurst, Robyn. 2001. *Bodies: Exploring Fluid Boundaries*. London: Routledge.

Lovering, John, and Hade Türkmen. 2011. "Bulldozer Neo-Liberalism in Istanbul: The State-Led Construction of Property Markets, and the Displacement of the Urban Poor." *International Planning Studies* 16 (1): 73–96.

Low, Nicholas P., and Brendan Gleeson. 1998. *Justice, Society and Nature: An Exploration of Political Ecology*. London: Routledge.

MacDonald, Alex. 2020. "Turkey's Mount Ida: The Frontline between Mining Giants and Local People." Middle East Eye. 9 October 2020. http://www .middleeasteye.net/news/turkey-mount-ida-environment-battleground.

MacGregor, Sherilyn. 2017. *Routledge Handbook of Gender and Environment*. London: Routledge.

Malpas, Jeff. 2018. *Place and Experience: A Philosophical Topography*, 2nd ed. London: Routledge.

Mann, Michael. 1984. "The Autonomous Power of the State: Its Origins, Mechanisms and Results." *European Journal of Sociology/Archives Européennes de Sociologie* 25 (2): 185–213.

Manzo, Lynne C., and Patrick Devine-Wrigth, eds. 2014. *Place Attachment Advances in Theory, Methods, and Applications*. London: Routledge.

Martin, Adrian, Nicole Gross-Camp, Bereket Kebede, Shawn McGuire, and Joseph Munyarukaza. 2014. "Whose Environmental Justice? Exploring Local and Global Perspectives in a Payments for Ecosystem Services Scheme in Rwanda." *Geoforum* 54: 167–77.

Martinez-Alier, J., L. Temper, D. Del Bene, and A. Scheidel. 2016. "Is There a Global Environmental Justice Movement?" *Journal of Peasant Studies* 43 (3): 731–55.

Martinez-Alier, Joan. 1995. "The Environment as a Luxury Good or 'Too Poor to Be Green'?" *Ecological Economics* 13 (1): 1–10.

———. 2002. *The Environmentalism of the Poor: A Study of Ecological Conflicts and Valuation.* Northampton, MA: Edward Elgar.

Mascarenhas, Michael. 2007. "Where the Waters Divide: First Nations, Tainted Water and Environmental Justice in Canada." *Local Environment* 12 (6): 565–77.

Massey, Doreen B. 1994. *Space, Place, and Gender.* Minneapolis: University of Minnesota Press.

Massumi, Brian. 2002. *Parables for the Virtual: Movement, Affect, Sensation.* Durham, NC: Duke University Press.

McAdam, Douglas. 1982. *Political Process and the Development of Black Insurgency.* Chicago: University of Chicago Press.

McCully, Patrick. 2001. *Silenced Rivers: The Ecology and Politics of Large Dams,* enlarged and updated edition. London: Zed Books.

McGregor, Deborah. 2015. "Indigenous Women, Water Justice and Zaagidowin (Love)." *Canadian Woman Studies* 30 (2–3): 71–78.

McMurtry, John. 2001. "The Life-Ground, the Civil Commons and the Corporate Male Gang." *Canadian Journal of Development Studies/Revue Canadienne d'études Du Développement* 22 (4): 819–54.

Mead, George H. 1934. *Mind, Self & Society.* Chicago: University of Chicago Press.

Mellor, Mary. 2003. "Gender and Environment." In *Ecofeminism and Globalization: Exploring Culture, Context, and Religion,* by H. Eaton and L. A. Lorentzen. Lanham, MD: Rowman & Littlefield.

Merchant, Carolyn. 1980. *The Death of Nature: Women, Ecology, and the Scientific Revolution.* San Francisco: Harper & Row.

Merleau-Ponty, Maurice. 1964. *The Primacy of Perception, and Other Essays on Phenomenological Psychology, the Philosophy of Art, History, and Politics.* Evanston, IL: Northwestern University Press.

———. 1968. *The Visible and the Invisible.* Evanston, IL: Northwestern University Press.

———. 2003. *Nature: Course Notes from the Collège de France.* Evanston, IL: Northwestern University Press.

———. 2012. *Phenomenology of Perception.* London: Routledge.

Merrick, Helen. 2017. "Naturecultures and Feminist Materialism." In *Routledge Handbook of Gender and Environment,* edited by Sherilyn MacGregor, 101–14. London: Routledge.

Mezzadra, Sandro, and Brett Neilson. 2019. *The Politics of Operations: Excavating Contemporary Capitalism.* Durham: Duke University Press.

Mies, Maria, and Vandana Shiva. 1993. *Ecofeminism.* London: Zed Books.

Milliyet. 2015. "Bakan: HES'ler Son Derece Çevreci [Minister Eroğlu: HPPs Are Highly Environmentalist]", 5 December 2015. http://www.milliyet.com .tr/bakan-eroglu-hes-ler-son-derece-cevreci-ankara-yerelhaber-778043/.

Milton, Kay. 2002. *Loving Nature: Towards an Ecology of Emotion.* London: Routledge.

Mitchell, Timothy. 2002. *Rule of Experts: Egypt, Techno-Politics, Modernity.* Berkeley: University of California Press.

Mohai, P., D. Pellow, and J. T. Roberts. 2009. "Environmental Justice." *Review of Environment and Resources* 34: 405–30.

Moi, Toril. 2001. *What Is a Woman?: And Other Essays.* Oxford: Oxford University Press.

Moore, Barrington. 1978. *Injustice: The Social Bases of Obedience and Revolt.* London: MacMillan.

Moore, Jason W. 2015. *Capitalism in the Web of Life: Ecology and the Accumulation of Capital.* London: Verso.

———. 2016. *Anthropocene or Capitalocene? Nature, History, and the Crisis of Capitalism.* Oakland, CA: PM Press.

Moran, Dermot. 2015. "Ineinandersein and L'interlacs: The Constitution of the Social World or 'We-World'(*Wir-Welt*) in Edmund Husserl and Maurice Merleau-Ponty." In *Phenomenology of Sociality: Discovering the "We,"* edited by Thomas Szanto and Dermot Moran, 107–26. London: Routledge.

Münch, Peter. 2020. "Ein Tag am Meer [One Day at Sea]." *Süddeutsche Zeitung,* 11 September 2020. https://www.sueddeutsche.de/panorama/israel-westjordanland-1.5027731.

Muraca, Barbara. 2016. "Relational Values: A Whiteheadian Alternative for Environmental Philosophy and Global Environmental Justice." *Balkan Journal of Philosophy* 8 (1): 19–38.

Murdoch, Jonathan. 1997. "Inhuman/Nonhuman/Human: Actor-Network Theory and the Prospects for a Nondualistic and Symmetrical Perspective on Nature and Society." *Environment and Planning D: Society and Space* 15 (6): 731–56.

Murton, Brain. 2012. "Being in the Place World: Toward a Māori 'Geographical Self.'" *Journal of Cultural Geography* 29 (1): 87–104.

Naess, Arne. 1989. *Ecology, Community, and Lifestyle.* Cambridge: Cambridge University Press.

Nagel, Thomas. 1986. *The View from Nowhere.* Oxford: Oxford University Press.

Nancy, Jean-Luc. 2000. *Being Singular Plural.* Stanford, CA: Stanford University Press.

Naples, Nancy A. 1998. *Community Activism and Feminist Politics: Organizing Across Race, Class, and Gender.* London: Routledge.

Nefes, Türkay Salim. 2018. "The Sociological Foundations of Turkish Nationalism." *Journal of Balkan and Near Eastern Studies* 20 (1): 15–30.

Neimanis, Astrida. 2017. *Bodies of Water: Posthuman Feminist Phenomenology.* London: Bloomsbury.

Neville, Kate J., and Glen Coulthard. 2019. "Transformative Water Relations:
Indigenous Interventions in Global Political Economies." *Global Environmental Politics* 19 (3): 1–15.

Nichols, Wallace J. 2014. *Blue Mind.* New York: Little, Brown.

Nightingale, Andrea. 2006. "The Nature of Gender: Work, Gender, and Environment." *Environment and Planning D: Society and Space* 24 (2): 165–85.

———. 2019. "Commoning for Inclusion? Commons, Exclusion, Property and
Socio-Natural Becomings." *International Journal of the Commons* 13 (1),
16–35.

Nixon, Rob. 2011. *Slow Violence and the Environmentalism of the Poor.*
Cambridge, MA: Harvard University Press.

Norgaard, Kari Marie, and Ron Reed. 2017. "Emotional Impacts of Environmental Decline: What Can Native Cosmologies Teach Sociology about
Emotions and Environmental Justice?" *Theory and Society* 46 (6):
463–95.

Novotny, Patrick. 2000. *Where We Live, Work, and Play: The Environmental
Justice Movement and the Struggle for a New Environmentalism.* Westport,
CT.: Praeger.

Nussbaum, Martha C. 2011. *Creating Capabilities: The Human Development
Approach.* Cambridge, MA: Belknap Press of Harvard University Press.

Oğuz, Sıla P. 2016. "Trabzon Solaklı Vadisi: Ogene Halklarının HES'lere Karşı
Mücadelesi [Solakli Valley of Trabzon: The Struggle of Ogene People against
the Hydropower Plants]." In *In Sudan Sebepler: Türkiye'de Neoliberal
Su-Enerji Politikaları ve Direnişler [Neoliberal Politics of Water-Energy in
Turkey and Resistance],* edited by Cemil Aksu, Sinan Erensü, and Erdem
Evren, 199–222. İstanbul: İletişim.

Oksala, Johanna. 2006. "Female Freedom: Can the Lived Body Be Emanci-
pated?" In *Feminist Interpretations of Maurice Merleau-Ponty,* edited by
D. E. Olkowski and G. Weiss, 209–28. University Park: Pennsylvania State
University Press.

Olkowski, Dorothea E. 2006. "Introduction: The Situated Subject." In *Feminist
Interpretations of Maurice Merleau-Ponty,* edited by D. E. Olkowski and G.
Weiss. University Park: Pennsylvania State University Press.

———. 2017. "Using Our Intuition: Creating the Future Phenomenological
Plane of Thought." In *Feminist Phenomenology Futures,* edited by H. A.
Fielding and D. E. Olkowski, 3–20. Bloomington: Indiana University
Press.

Olkowski, Dorothea E., and Geil Weiss. 2006. *Feminist Interpretations of Maurice Merleau-Ponty*. University Park: Pennsylvania State University Press.

Ollman, Bertell. 2003. *Dance of the Dialectic: Steps in Marx's Method*. Urbana: University of Illinois Press.

O'Loughlin, Marjorie. 1995. "Intelligent Bodies and Ecological Subjectivities: Merleau-Ponty's Corrective to Postmodernism's 'Subjects' of Education." In *Philosophy of Education Yearbook*, 1–6. Sydney: University of Sydney.

Oppermann, Serpil. 2013. "Feminist Ecocriticism: A Posthumanist Direction in Ecocritical Trajectory." In *International Perspectives in Feminist Ecocriticism*, edited by G. Gaard, S. C. Estok, and S. Oppermann. London: Routledge.

Ostrom, Elinor. 1990. *Governing the Commons: The Evolution of Institutions for Collective Action*. Cambridge: Cambridge University Press.

Ozan, Özge. 2010. "Ardanuçlı Kadınlar: Suyumuzu Satanlar Cenazelerini Yıkayacak Su Bulamasınlar [The Women of Ardanuç: The Ones Who Sold Our Water May Not Find Water to Wash Their Corpses]." Sendika.org, 24 July 2010. https://sendika63.org/2010/07/ardanuclu-kadinlar-suyumuzu-satanlar-cenazelerini-yikayacak-su-bulamasinlar-ozge-ozan-44715/.

Özbudun, Ergun. 2015. "Turkey's Judiciary and the Drift toward Competitive Authoritarianism." *International Spectator* 50 (2): 42–55.

Öztürk, Murat, Joost Jongerden, and Andy Hilton. 2018. "The (Re) Production of the New Peasantry in Turkey." *Journal of Rural Studies* 61: 244–54.

———. 2021. "Agriculture and Rural Life in Turkey." In *The Routledge Handbook on Contemporary Turkey*, edited by Joost Jongerden, 322–34. London: Routledge.

Özyürek, Esra, Gaye Özpınar, and Emrah Altındiş, eds. 2019. *Authoritarianism and Resistance in Turkey: Conversations on Democratic and Social Challenges*. Cham: Springer.

Pardo, Mary. 1990. "Mexican American Women Grassroots Community Activists: 'Mothers of East Los Angeles.'" *Frontiers: A Journal of Women Studies* 11 (1): 1–7.

Parris, Christie L., Karen A. Hegtvedt, Lesley A. Watson, and Cathryn Johnson. 2014. "Justice for All? Factors Affecting Perceptions of Environmental and Ecological Injustice." *Social Justice Research* 27 (1): 67–98.

Petersen, Felix, and Zeynep Yanaşmayan. 2020. *The Failure of Popular Constitution Making in Turkey: Regressing towards Constitutional Autocracy*. Cambridge: Cambridge University Press.

Petras, James. 2013. "Brazil: Extractive Capitalism and the Great Leap Backward." *World Review of Political Economy* 4 (4): 469–83.

Philips, Mary, and Nick Rumens, eds. 2015. *Contemporary Perspectives on Ecofeminism*. Routledge Explorations in Environmental Studies. London: Routledge.

Piazza, Gianni, and Donatella Della Porta. 2008. *Voices of the Valley, Voices of the Straits: How Protest Creates Communities.* New York: Berghahn Books.

Pile, Steve. 2010. "Emotions and Affect in Recent Human Geography." *Transactions of the Institute of British Geographers* 35 (1): 5–20.

Plant, J. 1989. *Healing the Wounds: The Promise of Ecofeminism.* Philadelphia: New Society.

Plumwood, Val. 1993. *Feminism and the Mastery of Nature.* London: Routledge.

Polanyi, Karl. 2014. *The Great Transformation: The Political and Economic Origins of Our Time.* Boston: Beacon Press.

Porritt, Jonathon. 1984. *Seeing Green: The Politics of Ecology Explained.* Oxford: Blackwell.

Posey, Darrell Addison, and United Nations Environment Programme, eds. 1999. *Cultural and Spiritual Values of Biodiversity.* London: Intermediate Technology.

Prindeville, D. M. 2004. *On the Streets and in the State House: American Indian and Hispanic Women and Environmental Policymaking in New Mexico.* London: Routledge.

Proshansky, Harold M. 1978. "The City and Self-Identity." *Environment and Behavior* 10 (2): 147–69.

Radikal. 2008. "Erdoğan Rize'de Santralleri Savundu [Erdoğan Defended the Powerplants in Rize], "22 August 2008.

———. 2015. "Havva Ana Yalnız Değil: Çevreyi Paçavra Ettunuz . . . Biz de Sizi Edeceğuz [Mother Havva (Eve) Is Not Alone: You Have Messed with the Environment . . . We Will Mess with You]," 19 July 2015.

Ramos, Howard. 2015. "Mapping the Field of Environmental Justice: Redistribution, Recognition and Representation in ENGO Press Advocacy." *Canadian Journal of Sociology* 40 (3): 355–76.

Rawls, John. 1971. *A Theory of Justice.* Cambridge, MA: Belknap Press of Harvard University Press.

Raynes, Dakota K. T., Tamara L. Mix, Angela Spotts, and Ariel Ross. 2016. "An Emotional Landscape of Place-Based Activism: Exploring the Dynamics of Place and Emotion in Antifracking Actions." *Humanity & Society* 40 (4): 401–23.

Resurrección, Bernadette P. 2017. "Gender and Environment in the Global South: From 'Women, Environment and Development' to Feminist Political Ecology." In *Routledge Handbook of Gender and Environment,* edited by Sherilyn MacGregor, 71–85. London: Routledge.

Ribeiro, G. L. 1994. *Transnational Capitalism and Hydropolitics in Argentina: The Yacyretá High Dam.* Gainesville: University Press of Florida.

Rich, Adrienne. 1984. "Notes Towards a Politics of Location." In *Blood, Bread and Poetry: Selected Prose 1979–1985,* edited by Adrienne Rich. London: Little Brown.

Ricoveri, Giovanna. 2013. *Nature for Sale: The Commons versus Commodities.* London: Pluto Press.

Rocheleau, D., B. Thomas-Slayter, and E. Wangari. 1996. *Feminist Political Ecology: Global Issues and Local Experiences.* London: Routledge.

Rosa, Hartmut. 2019. *Resonance: A Sociology of Our Relationship to the World.* Cambridge: Polity Press.

Rose, Mitch. 2012. "Dwelling as Marking and Claiming." *Environment and Planning D: Society and Space* 30 (5): 757–71.

Rose, Mitch, and John Wylie. 2006. *Animating Landscape.* London: SAGE.

Rosiek, Jerry Lee, Jimmy Snyder, and Scott L. Pratt. 2020. "The New Materialisms and Indigenous Theories of Non-Human Agency: Making the Case for Respectful Anti-Colonial Engagement." *Qualitative Inquiry* 26 (3–4): 331–46.

Rusansky, Tamara. 2020. "Embroidering Resistance: Daily Struggles of Women Affected by the Baixo Iguaçu Hydropower Dam in Paraná, South Brazil." ISS Working Paper Series / General Series (Vol. 654). International Institute of Social Studies of Erasmus University (ISS). https://repub.eur.nl/pub /124860/.

Sack, Robert David. 1997. *Homo Geographicus: A Framework for Action, Awareness, and Moral Concern.* Baltimore: Johns Hopkins University Press.

Şahinde, Yavuz, and Özlem Şendeniz. 2013. "HES Direnişlerinde Kadınların Deneyimleri: Fındıklı Örneği." *Fe Dergi* 5 (1): 43–58.

Salamon, Gayle. 2018. "What's Critical about Critical Phenomenology?" *Journal of Critical Phenomenology* 1: 8–17.

Samaddar, Ranabir. 2007. *The Materiality of Politics,* Volume 2. London: Anthem Press.

Sasson-Levy, Orna, and Tamar Rapoport. 2003. "Body, Gender, and Knowledge in Protest Movements: The Israeli Case." *Gender & Society* 17 (3): 379–403.

Schlosberg, David. 2004. "Reconceiving Environmental Justice: Global Movements and Political Theories." *Environmental Politics* 13 (3): 517–40.

———. 2007. *Defining Environmental Justice: Theories, Movements, and Nature.* Oxford: Oxford University Press.

———. 2013. "Theorising Environmental Justice: The Expanding Sphere of a Discourse." *Environmental Politics* 22 (1): 37–55.

Schutz, Alfred, and Thomas Luckmann. 1973. *The Structures of the Life-World.* Evanston, IL: Northwestern University Press.

Scott, Joan W. 1986. "Gender: A Useful Category of Historical Analysis." *American Historical Review* 91 (5): 1053–75.

———. 1991. "The Evidence of Experience." *Critical Inquiry* 17 (4): 773–97.

———. 2010. "Gender: Still a Useful Category of Analysis?" *Diogenes* 57 (1): 7–14.

Seabrook, Liz. 2020. "Swimming with Selkies." *Suitcase Magazine* 30 (The Health Issue): 140–49.

Seager, Joni. 1993. *Earth Follies: Feminism, Politics and the Environment.* London: Earthscan.

———. 1996. "Hysterical Housewives and Other Mad Women." In *Feminist Political Ecology: Global Issues and Local Experiences,* edited by D. Rocheleau, B. Thomas-Slayter, and E. Wangari, 271–83. London: Routledge.

Seamon, David. 1979. *A Geography of the Lifeworld: Movement, Rest and Encounter.* London: Croom Helm.

Şekercioğlu, Çağan H., Sean Anderson, Erol Akçay, Raşit Bilgin, Özgün Emre Can, Gürkan Semiz, Çağatay Tavşanoğlu, Mehmet Baki Yokeş, Anıl Soyumert, and Kahraman Ipekdal. 2011. "Turkey's Globally Important Biodiversity in Crisis." *Biological Conservation* 144 (12): 2752–69.

Sen, Amartya. 2009. *The Idea of Justice.* Cambridge, MA: Belknap Press of Harvard University Press.

Şendeniz, Özlem, and Emek Yıldırım. 2018. *Sırtında Sepeti: Bafra'dan Hopa'ya Karadeniz'de Kadınlık Halleri [Basket on the Back: Womanhood in the Black Sea from Bafra to Hopa].* Ankara: Phoenix.

Shiva, Vandana. 1989. *Staying Alive: Women, Ecology and Development.* London: Zed Books.

Simonian, Hovann H. 2007. *The Hemshin: History, Society and Identity in the Highlands of Northeast Turkey.* London: Routledge.

Simonsen, Kirsten. 2012. "In Quest of a New Humanism: Embodiment, Experience and Phenomenology as Critical Geography." *Progress in Human Geography* 37 (1): 10–26.

Simonsen, Kirsten, and Lasse Koefoed. 2020. *Geographies of Embodiment: Critical Phenomenology and the World of Strangers.* London: SAGE.

Singh, Neera M. 2013. "The Affective Labor of Growing Forests and the Becoming of Environmental Subjects: Rethinking Environmentality in Odisha, India." *Geoforum* 47: 189–98.

Skinner, Jonathan. 2012. *The Interview: An Ethnographic Approach.* London: Berg.

Sneddon, Chris, and Coleen Fox. 2011. "The Cold War, the US Bureau of Reclamation, and the Technopolitics of River Basin Development, 1950–1970." *Political Geography* 30 (8): 450–60.

Sneddon, Chris, Leila Harris, Radoslav Dimitrov, and Özsemi Uygar. 2002. "Contested Waters: Conflict, Scale, and Sustainability in Aquatic Socioecological Systems." *Society & Natural Resources* 15 (8): 663–75.

Snow, David A., and Danny Trom. 2002. "The Case Study and the Study of Social Movements." In *Methods of Social Movement Research,* edited by Bert Klandermans and Suzanne Staggenborg, Volume 16, 146–72. Minneapolis: University of Minnesota Press.

Soja, Edward W. 1996. *Thirdspace: Journeys to Los Angeles and Other Real-and-Imagined Places.* Malden, MA: Blackwell.

214   REFERENCES

Sözen, Ülker. 2019. "Culture, Politics and Contested Identity among the 'Kurdish' Alevis of Dersim: The Case of the Munzur Culture and Nature Festival." *Journal of Ethnic and Cultural Studies* 6 (1): 63–76.

Sputnik. 2018. "Arhavi'de 'Çılgın HES' Deneme Retimine Geçti, Dere Kurudu [The Crazy HEPP Started Test Production, the River Has Dried Up]," 21 September 2018. https://tr.sputniknews.com/cevre/201809211035311601-arhavi-hes-dere-cilgin/.

Stawarska, Beata. 2006. "From the Body Proper to Flesh: Merleau-Ponty on Intersubjectivity." In *Feminist Interpretations of Maurice Merleau-Ponty*, edited by D. Olkowski and G. Weiss, 91–106. University Park: Pennsylvania State University Press.

Stedman, Richard C. 2002. "Toward a Social Psychology of Place: Predicting Behavior from Place-Based Cognitions, Attitude, and Identity." *Environment and Behavior* 34 (5): 561–81.

Stefanovic, Ingrid Leman. 2020a. "Introduction." In *The Wonder of Water: Lived Experience, Policy and Practice*, edited by I. L. Stefanovic, 3–8. Toronto: Toronto University Press.

———. 2020b. "Water and the City: Towards an Ethos of Fluid Urbanism." In *The Wonder of Water: Lived Experience, Policy and Practice*, edited by I. L. Stefanovic, 114–32. Toronto: Toronto University Press.

Stein, Rachel, ed. 2004. *New Perspectives on Environmental Justice: Gender, Sexuality, and Activism*. New Brunswick, NJ: Rutgers University Press.

Steward, Susan. 2005. "Remembering the Senses." In *Empire of the Senses: The Sensual Culture Reader*, edited by D. Howes, 59–69. Oxford: Berg.

STHP (Suyun Ticarileşmesine Hayır Platformu) (No to Commercialization of Water Platform). 2012. "Su Kanunu Tasarısı Üzerine Yapılan Çalıştay Sonuç Bildirgesi (Conclusion of the Workshop on the Draft Water Law)." 2012. http://www.supolitik.org/su_kanun_tasarisi_calistayi_STHP.htm.

Stoller, Silvia. 2000. "Reflections on Feminist Merleau-Ponty Skepticism." *Hypatia* 15 (1): 175–82.

Strang, Veronica. 2004. *The Meaning of Water*. Oxford: Berg.

———. 2005. "Common Senses: Water, Sensory Experience and the Generation of Meaning." *Journal of Material Culture* 10 (1): 92–120.

———. 2009. *Gardening the World*. Oxford: Berghahn Books.

———. 2014. "Fluid Consistencies: Material Relationality in Human Engagements with Water." *Archaeological Dialogues* 21 (2): 133–50.

———. 2015. *Water: Nature and Culture*. London: Reaktion Books.

Sturgeon, Noël. 1997. *Ecofeminist Natures: Race, Gender, Feminist Theory and Political Action*. London: Routledge.

Su Hakkı Kampanyası (Right to Water Campaign). 2016. "Susmuyor; suyumuzu, su hakkımızı, yaşamı savunuyoruz!" https://www.suhakki.org/basin-aciklamalari/susmuyor-suyumuzu-su-hakkimizi-yasami-savunuyoruz/.

Sultana, Farhana. 2009. "Fluid Lives: Subjectivities, Gender and Water in Rural Bangladesh." *Gender, Place and Culture* 16 (4): 427–44.

Sultana, Farhana, and Alex Loftus. 2013. *The Right to Water: Politics, Governance and Social Struggles.* Hoboken, NJ: Taylor & Francis.

Sumner, J. 2005. *Sustainability and the Civil Commons: Rural Communities in the Age of Globalization.* Toronto: University of Toronto Press.

Surrallés, Alexandre, and Pedro García Hierro. 2005. *The Land Within: Indigenous Territory and the Perception of Environment.* Copenhagen: IWGIA (International Work Group for Indigenous Affairs).

Svarstad, Hanne, and Tor A. Benjaminsen. 2020. "Reading Radical Environmental Justice through a Political Ecology Lens." *Geoforum* 108: 1–11.

Swyngedouw, Erik. 1999. "Modernity and Hybridity: Nature, Regeneracionismo, and the Production of the Spanish Waterscape, 1890–1930." *Annals of the Association of American Geographers* 89 (3): 443–65.

———. 2015. *Liquid Power: Contested Hydro-Modernities in Twentieth- Century Spain.* Cambridge, MA: MIT Press.

Sylvain, Renée. 2002. "'Land, Water, and Truth': San Identity and Global Indigenism." *American Anthropologist* 104 (4): 1074–85.

Taylor, Charles. 1992. *Mutliculturalism and "the Politics of Recognition."* Princeton, NJ: Princeton University Press.

Taylor, Verta. 1998. "Feminist Methodology in Social Movements Research." *Qualitative Sociology* 21 (4): 357–79.

———. 2000. "Mobilizing for Change in a Social Movement Society." *Contemporary Sociology* 29 (1): 219–30.

Thomas, Amanda C. 2015. "Indigenous More-than-Humanisms: Relational Ethics with the Hurunui River in Aotearoa New Zealand." *Social & Cultural Geography* 16 (8): 974–90.

Thomas, Charis, and Sherilyn MacGregor. 2017. "The Death of Nature: Foundations of Ecological Feminist Thought." In *Routledge Handbook of Gender and Environment*, 43–53. London: Routledge.

Thrift, Nigel J. 2008. *Non-Representational Theory: Space, Politics, Affect.* London: Routledge.

Tilly, Christopher. 2004. *The Materiality of Stone: Explorations in Landscape Phenomenology*, 1st ed. Oxford: Berg.

Tilt, B. 2015. *Dams and Development in China: The Moral Economy of Water and Power.* New York: Colombia University Press.

TMMOB (Union of Chambers of Turkish Engineers and Architects). 2011. "Hidroelektrik Santaller Raporu [Report on the Hydroelectric Plants]." TMMOB—Türk Mühendis ve Mimar Odaları Birliği. http://www.tmmob.org.tr /sites/default/files/682384b57999789_ek.pdf.

Toadvine, Ted. 2009. *Merleau-Ponty's Philosophy of Nature.* Evanston, IL: Northwestern University Press.

———. 2015. "Phenomenology and Environmental Ethics." In *The Oxford Handbook of Environmental Ethics,* edited by S. M. Gardiner and Allen Thompson, 174–85. Oxford: Oxford University Press.

Toktamış, Kumru F., and Isabel David. 2018. "Introduction: Democratization Betrayed—Erdogan's New Turkey." *Mediterranean Quarterly* 29 (3): 3–10.

Tsing, Anna Lowenhaupt. 2015. *The Mushroom at the End of the World: On the Possibility of Life in Capitalist Ruins.* Princeton, NJ: Princeton University Press.

Turhan, Ethemcan, and Arif Cem Gündoğan. 2019. "Price and Prejudice: The Politics of Carbon Market Establishment in Turkey." *Turkish Studies* 20 (4): 512–40.

Turkish Water Assembly. 2010. "Water Manifest of the Turkish Water Assembly." Turkish Water Assembly. http://turkishwaterassembly.weebly.com /uploads/6/7/0/8/6708484/watermanifestoftheturkishwaterassembly.pdf.

———. 2011. "HEPP's, Dams and the Status of Nature in Turkey." Studylib. https://studylib.net/doc/7537045/hepp-s—dams-and-the-status-of-nature-in-turkey-turkish-w . . .

Turner, R. H., and L. M. Killian. 1987. *Collective Behavio,* 3rd ed. Englewood Cliffs, NJ: Prentice-Hall.

Uğurlu, Sinem. 2015. "Arhavi'de Direnen Köylüler: Dere Kurudu Mu Biz de Kuruduk Demektir [Villagers Resisting in Arhavi: When the River Dries Out It Means We Dry Out, Too]." Evrensel, 26 June 2015. https://www .evrensel.net/haber/254548/arhavide-hese-karsi-direnen-koyluler-dere-kurudu-mu-biz-de-kuruduk-demektir.

Ulloa, Astrid. 2017. "Perspectives of Environmental Justice from Indigenous Peoples of Latin America: A Relational Indigenous Environmental Justice." *Environmental Justice* 10 (6): 175–80.

UN. 2010. "General Assembly Res/64/292. The Human Right to Water and Sanitation." United Nations. https://www.un.org/en/ga/search/view_doc .asp?symbol=A/RES/64/292.

Ünal, Burcu. 2013. "Karadeniz Direniyor [The Black Sea Is Resisting]." *Milliyet,* 7 February 2013. https://www.milliyet.com.tr/gundem /karadeniz-direniyor-burada-hes-deyince-akan-sular-duruyor-1730744.

Unger, Nancy C. 2012. *Beyond Nature's Housekeepers: American Women in Environmental History.* Oxford: Oxford University Press.

Vakoch, Douglas A., and Sam Mickey. 2017. *Women and Nature? Beyond Dualism in Gender, Body, and Environment.* London: Routledge.

Velicu, Irina. 2015. "Demonizing the Sensible and the 'Revolution of Our Generation' in Rosia Montana." *Globalizations* 12 (6): 846–58.

Verchick, Robert, R. M. 2004. "Feminist Theory and Environmental Justice." In *New Perspectives on Environmental Justice: Gender, Sexuality, and Activism,* edited by R. Stein, 63–77. New Brunswick, NJ: Rutgers University Press.

Verhoeven, Harry. 2018. *Environmental Politics in the Middle East*. Oxford: Oxford University Press. https://public.ebookcentral.proquest.com/choice /publicfullrecord.aspx?p=5558541.

Vermeylen, Saskia, and Gordon Walker. 2011. "Environmental Justice, Values and Biological Diversity: The San and the Hoodia Benefit Sharing Agreement." In *Environmental Inequalities Beyond Borders: Local Perspectives on Global Injustices*, edited by J. Carmin and J. Angyeman, 105–28. Cambridge, MA: MIT Press.

Walker, Gordon. 2009. "Beyond Distribution and Proximity: Exploring the Multiple Spatialities of Environmental Justice." *Antipode* 41 (4): 614–36.

———. 2012. *Environmental Justice: Concepts, Evidence and Politics*. London: Routledge.

Walsh, Casey. 2018. *Virtuous Waters: Mineral Springs. Bathing and Infrastructure in Mexico*. Oakland: University of California Press.

Walsh, Edward J. 1981. "Resource Mobilization and Citizen Protest in Communities around Three Mile Island." *Social Problems* 29 (1): 1–21.

Ward, Christopher, and Sandra Ruckstuhl. 2017. *Water Scarcity, Climate Change and Conflict in the Middle East: Securing Livelihoods, Building Peace*. London: I. B. Tauris.

Waster-Herber, Misse. 2004. "Underlying Concerns in Land-Use Conflicts: The Role of Place-Identity in Risk Perception." *Environmental Science & Policy* 7 (2): 109–16.

Webber, Jeffery R. 2015. "The Indigenous Community as 'Living Organism': José Carlos Mariátegui, Romantic Marxism, and Extractive Capitalism in the Andes." *Theory and Society* 44 (6): 575–98.

Weber, Andreas. 2016. *Biopoetics: Towards an Existential Ecology*. New York: Springer.

Weiss, Gail. 1999. *Body Images Embodiment as Intercorporeality*. London: Routledge.

Weiss, Gail, Ann V. Murphy, and Gayle Salamon. 2019. *50 Concepts for a Critical Phenomenology*. Evanston, IL: Northwestern University Press.

Whatmore, Sarah. 2002. *Hybrid Geographies: Natures Cultures Spaces*. London: SAGE.

White, David, and Marc Herzog. 2016. "Examining State Capacity in the Context of Electoral Authoritarianism, Regime Formation and Consolidation in Russia and Turkey." *Southeast European and Black Sea Studies* 16 (4): 551–69.

Whitmore, Luke. 2018. *Mountain, Water, Rock, God*. Oakland: University of California Press.

Whyte, Kyle Powys. 2011. "The Recognition Dimensions of Environmental Justice in Indian Country." *Environmental Justice* 4 (4): 199–205.

Wilson, Nicole J., and Jody Inkster. 2018. "Respecting Water: Indigenous Water Governance, Ontologies, and the Politics of Kinship on the Ground." *Environment and Planning E: Nature and Space* 1 (4): 516-38.

Woods, Michael, Jon Anderson, Steven Guilbert, and Suzie Watkin. 2012. "'The Country (Side) Is Angry': Emotion and Explanation in Protest Mobilization." *Social & Cultural Geography* 13 (6): 567–85.

World Bank. 2011. "Clean Technology Fund Investments in Turkey: Comment on Berne Declaration Report." Text/HTML. World Bank. 12 October 2011. https://doi.org/10/12/clean-technology-fund-investments-in-turkey-comment-on-berne-declaration-report.

WWF. 2013. "Doğa Korumaya Yasayla Darbe [Legal Attack Against Conservation of Nature]." World Wide Fund for Nature. https://www.wwf.org.tr/?1354.

Wylie, John. 2003. "Landscape, Performance and Dwelling: A Glastonbury Case Study." In *Country Visions,* edited by Paul Clocke, 136–57. Harlow, UK: Pearson.

Ya-Chu Yang, Karen. 2017. "Introduction." In *Women and Nature? Beyond Dualism in Gender, Body, and Environment,* edited by Douglas A. Vakoch and Sam Mickey, 3–9. London: Routledge.

Yackley, Ayla Jean. 2020. "Flooding in Turkey Kills at Least Eight in Bitter Reminder of Climate Chaos." Al-Monitor. 24 August 2020. https://www.al-monitor.com/pulse/originals/2020/08/flooding-kills-eight-turkey.html.

Yaka, Özge. 2017. "A Feminist-Phenomenology of Women's Activism against Hydropower Plants in Turkey's Eastern Black Sea Region." *Gender, Place and Culture* 24 (6): 868–89.

———. 2019a. "Gender and Framing: Gender as a Main Determinant of Frame Variation in Turkey's Anti-Hydropower Movement." *Women's Studies International Forum* 74: 154–61.

———. 2019b. "Rethinking Justice: Struggles for Environmental Commons and the Notion of Socio-Ecological Justice." *Antipode* 51 (1): 353–72.

———. 2020. "Justice as Relationality: Socio-Ecological Justice in the Context of Anti-Hydropower Movements in Turkey." *DIE ERDE–Journal of the Geographical Society of Berlin* 151 (2–3): 167–80.

Yaka, Özge, and Serhat Karakayali. 2017. "Emergent Infrastructures: Solidarity, Spontaneity and Encounter at Istanbul's Gezi Park Uprising." In *Protest Camps in International Context: Spaces, Infrastructures and Media of Resistance,* edited by Gavin Brown, Anna Feigenbaum, Fabian Frenzel, and Patrick McCurdy, 53–69. Bristol, UK: Policy Press.

Yaman, Melda. 2019. "Tarımsal Üretimde Kadın Emeği: Tarihte Kısa Bir Gezinti [Women's Labour in Agriculture: A Short History]." In *Aramızda Kalmasın: Kır, Kent ve Ötesinde Toplumsal Cinsiyet [Not to Keep Between Us: Gender in and Beyond the Country and the City],* edited by Özlem

Şendeniz. Fındıklı: Aramızda Toplumsal Cinsiyet Derneği Yayınları. https://
aramizda.org.tr/wp-content/uploads/2020/02/Aramızda_Kalmasın.pdf.

Yates, Julian S., Leila M. Harris, and Nicole J. Wilson. 2017. "Multiple Ontologies of Water: Politics, Conflict and Implications for Governance." *Society
and Space* 35 (5): 797–815.

Yavuz, Filiz. 2014. "Ahmetler Köyü HES Direnişi Sürüyor [The HEPP Resistance in the Ahmetler Village Continues]." *Cumhuriyet*, 2 September 2014.
https://www.cumhuriyet.com.tr/haber/ahmetler-koyu-hes-direnisini-
surduruyor-39449.

Yeğen, Mesut. 2007. "Turkish Nationalism and the Kurdish Question." *Ethnic
and Racial Studies* 30 (1): 119–51.

Yılmaz, Zafer. 2020. "Erdoğan's Presidential Regime and Strategic Legalism:
Turkish Democracy in the Twilight Zone." *South European Society and
Politics* 20 (2): 265–87.

Yılmaz, Zafer, and Bryan S. Turner. 2019. "Turkey's Deepening Authoritarianism and the Fall of Electoral Democracy." *British Journal of Middle Eastern
Studies* 4 (5): 691–698.

Young, Iris Marion. 1980. "Throwing Like a Girl: A Phenomenology of Feminine Body Comportment Motility and Spatiality." *Human Studies* 3 (1):
137–56.

———. 1990. *Justice and the Politics of Difference*. Princeton, NJ: Princeton
University Press.

———. 2002. "Lived Body vs Gender: Reflections on Social Structure and
Subjectivity." *Ratio* 15 (4): 410–28.

———. 2005. *On Female Body Experience: "Throwing Like a Girl" and Other
Essays*. New York: Oxford University Press.

Yuksek, Omer, Murat Ihsan Komurcu, Ibrahim Yuksel, and Kamil Kaygusuz.
2006. "The Role of Hydropower in Meeting Turkey's Electric Energy
Demand." *Energy Policy* 34 (17): 3093–103.

Yüksel, Ibrahim. 2013. "Renewable Energy Status of Electricity Generation and
Future Prospect Hydropower in Turkey." *Renewable Energy* 50: 1037–43.

Yüksel, Recai, and Ilker Zeren. 2010. " 'HES'lere neden karşı çıkılıyor,
anlamıyorum' ['I do not understand the opposition against HPPs']. " İhlas
Haber Ajansı (Ihlas News Agency). 2010. https://www.iha.com.tr/haber-heslere-
neden-karsi-cikiliyor-anlamiyorum-145274/.

Zürcher, Erik J. 1993. *Turkey: A modern history*. London: Tauris.

Index

Page numbers in *italics* denote figures.

narratives, 26; gendered, 52; tyrant state, 112
Nationalist Action Party (MHP), 38, 47; stronghold in East Black Sea Region (EBR), 38
"natural resources," 11, 177n23
Negri, Antonio, 27, 28
Neilson, Brett, 173n19
Neimanis, Astrida, 83–84, 98, 149, 181n9
Neville, Kate, 140–141
New Imperialism, The, 173n18
New Zealand, 148
Nichols, Wallace J., 82–83
Nixon, Rob, 11, 32, 79–80, 136, 137, 156
nonhumans, 63, 118
"nonproperty," 27
normativity, grounded, 133, 147
Nussbaum, Martha, 182n1

Öcalan, Abdullah, 37
Okumuşoğlu, Yakup, 147, 174–175n33
ontology, relational, 14, 75, 139–145, 164; and justice, 152, 155; Merleau-Ponty, Maurice, 93, 104, 139; and relational ethics, 146–149
Orme (dam), 126, 145
Oxa, Juan, 180n5
Ozan, Özge, 55, 60–61
Özlüer, Fevzi, 44, 174–175n33
Öztürk, Murat, 115

Paker, Hüseyin, 117
patriarchy, 77–81; classical, 63; negotiating, 77–78, 179n32; transformation from within, 77–78
pensée de survol, 73
perception: Merleau-Ponty, Maurice, 93; sensory, 93
Perception of the Environment, The, 143
phantom limb, 10, 124
phenomenology: critical, 76; feminist, 77, 160–164
PKK, 37, 47
place: attachment, 114, 118; being in, 110; identity and, 114, 118, 127; Indigenous conceptions of, 111, 118; "making of," 110, 111–112, 118–119; phenomenological conceptions of, 111; relational nature of, 109; sense of, 13–14, 110–111, 112, 116–117, 119, 126; thinned-out, 180n2
Polanyi, Karl, 29–30

"primitive accumulation," 29–30, 35
proletarianization, forced, 35
Proust, Marcel, 121, 125

Qing, Dai, 56

rational choice theory, 146
Rawls, John, 134, 182n1, 182n2. See also *Theory of Justice, A*
reduction, phenomenological, 10, 87–88
relationality: ethics, relational, 146–149; excess of, 155; gender aspects or, 9; human-nonhuman, 152; more-than-human, 146–149; ontology, 14, 75, 139–145, 164
research process, 2–5
Rew, Kate, 99
rivers: as legal persons, 148; as nonhuman persons, 137–138, 140, 141; Qu'ran on, 117
Rosa, Hartmut, 8, 11, 97–98, 104, 121, 139–140, 143

Samaddar, Ranabir, 133
Sarıyıldız, Özlem, 174n32
Sardar Sarovar, 22
Sarıca, Bora, 174–175n33
Scarboro, Lorelei, 126
Schlosberg, David, 183n6
Seabrook, Liz, 99
Semiha Teyze, 78–79
Sen, Amartya, 182n1
senses, 93; convergence of the, 101–102; hearing, 84, 91, 93, 119, 122; sight, 84, 89, 90, 93, 122, 159, 163; smell, 84, 89, 93; taste, 84, 93, 102, 119, 159; touch, 84, 93, 96, 100, 102, 103, 105, 119, 122, 124, 159
sensory: capacities, 73, 75, 84; experiences, 73, 77, 84
Simpson, Leanne Betasamosake, 147
skin, 97–98
"slow violence," 32
society-nature dualism, 149
Sönmez, Yücel, 56
South Africa, 138
Southeast Anatolia Project, 2–3
Southeast Anatolia Region, *4*
St. James Parish, 50, 62
Standing Rock, 50, 138
State Hydraulic Works (DSI), 23, 24
Stefanovic, Ingrid Leman, 82–83, 84, 86

stereotypes, strategical use of, 63
Strang, Veronica, 84, 89
struggles, antihydropower, 26, 26–27; chro-
  nology of, 3; conspiracy theories about,
  52; documentaries about, 26; in East and
  Southeast Anatolia Regions, 36–38; in
  East Black Sea Region (EBR), 38–39;
  gender differences and, 52ff.; with legal
  means, 42–43; in Mediterranean Region
  (MR), 35–36; and protection of livelihoods, 35–36; regional differences in,
  34–39; social, 42–43; tactics of, 41–42;
  transgressiveness of, 50; and War of Independence, 52
Su Hakki Kampanyasi (Water Rights Campaign), 27
Su Meclisi (Water Assembly), 39, 168n6
Sudaki Suretler, 26–27, 34, 90, 126
SuPolitik ("WaterPolitics"), 27
Susuz Yaz ("Dry Summer"), 26
Suyun Ticarileşmesine Hayır *Platformu (*No
  to the Commercialization of Water Platform), 39–40

tea (monocultural production), 63–64, 117
tea farming, 177n19
Theory of Justice, A, 182n2. *See also* Rawls,
  John
Three Gorges (dam), 22, 56
"throwing like a girl," 74
Tigris (river), 2–3
Toadvine, Ted, 103, 148–149
"transcorporeality," 101
Tsleil-Waututh Nation, 138
Tunceli (Dersim), 37–38
Turkish Electricity Market Act, 23, 24
*Turkish Foundation for Combating Soil Ero-
  sion* (TEMA), 40
Türkkan, Ahmet, 46
Türkmen, Hade, 168n2

Uexküll, Jakob Johann von, 16
Ulutaş, Berna Babaoğlu, 174–175n33
umwelt, 16
Undomesticated Ground, 97
Unger, Nancy, 62

unimagined communities, 32, 137
Union of Chambers of Turkish Engineers and
  Architects (TMMOB), 21, 168n6
Uygar, Kömert, 174–175n33

Visible and the Invisible, The, 96

Walsh, Casey, 84–85, 88
water: bodies, 83–84; commodification of,
  27; rights, 27; symbolic meaning of,
  83–84
water-as-lifeblood, 148
water-as-ressource, 148
Water Use Agreement Act (2003), 23
WED (women, environment, and development) approaches, 62–63
Weiss, Gail, 75, 170n18
welfare state, 136
West Black Sea Region (WBR), 4
Whanganui (river), 148. *See also* rivers as
  legal persons
Whitehead, A. N., 140
Williams, Raymond, 126–127
Wilson, Nicole J., 141, 148
women, 51; peasant, 50–51; political subjectivity of, 103; representational roles of, 78;
  ungovernable, 79–80; visibility of, 50–51
"Women Sparrowhawks" (*Kadın Atmacalar*),
  58
Woods, Michael, 106
World Bank, 168n4, 172n5

Yates, Julian S., 148
Yavapai people, 145
yayla, 117, 119, 122, 175n1
"yellow scarf" movement, 112
Yıldırım, Barış 174–175n33
Yılmaz, Sezgin, 46
Yörük, 109–110
Young, Iris Marion, 9, 74, 170n18,
  182–183n4
Yukon First Nations, 141
yurt, 109, 115
Yusufeli, 3

Zaza Alevi population, 37–38

Founded in 1893,
UNIVERSITY OF CALIFORNIA PRESS
publishes bold, progressive books and journals
on topics in the arts, humanities, social sciences,
and natural sciences—with a focus on social
justice issues—that inspire thought and action
among readers worldwide.

The UC PRESS FOUNDATION
raises funds to uphold the press's vital role
as an independent, nonprofit publisher, and
receives philanthropic support from a wide
range of individuals and institutions—and from
committed readers like you. To learn more, visit
ucpress.edu/supportus.